T0189358

Advances in Intelligent Systems and Computing

Volume 1063

The series "Advances in Intelligent Systems and Computing" contains publications on theory, applications, and design methods of Intelligent Systems and Intelligent Computing. Virtually all disciplines such as engineering, natural sciences, computer and information science, ICT, economics, business, e-commerce, environment, healthcare, life science are covered. The list of topics spans all the areas of modern intelligent systems and computing such as: computational intelligence, soft computing including neural networks, fuzzy systems, evolutionary computing and the fusion of these paradigms, social intelligence, ambient intelligence, computational neuroscience, artificial life, virtual worlds and society, cognitive science and systems, Perception and Vision, DNA and immune based systems, self-organizing and adaptive systems, e-Learning and teaching, human-centered and human-centric computing, recommender systems, intelligent control, robotics and mechatronics including human-machine teaming, knowledge-based paradigms, learning paradigms, machine ethics, intelligent data analysis, knowledge management, intelligent agents, intelligent decision making and support, intelligent network security, trust management, interactive entertainment, Web intelligence and multimedia.

The publications within "Advances in Intelligent Systems and Computing" are primarily proceedings of important conferences, symposia and congresses. They cover significant recent developments in the field, both of a foundational and applicable character. An important characteristic feature of the series is the short publication time and world-wide distribution. This permits a rapid and broad dissemination of research results.

** Indexing: The books of this series are submitted to ISI Proceedings, EI-Compendex, DBLP, SCOPUS, Google Scholar and Springerlink **

More information about this series at http://www.springer.com/series/11156

Joong Hoon Kim · Zong Woo Geem ·
Donghwi Jung · Do Guen Yoo · Anupam Yadav
Editors

Advances in Harmony Search, Soft Computing and Applications

 Springer

Editors
Joong Hoon Kim
School of Civil, Environmental
and Architectural Engineering
Korea University
Seoul, Korea (Republic of)

Zong Woo Geem
Department of Energy IT
Gachon University
Seongnam, Kwangju-jikhalsi
Korea (Republic of)

Donghwi Jung
Keimyung University
Daegu, Korea (Republic of)

Do Guen Yoo
Department of Civil Engineering
Suwon University
Gyeonggi-do, Korea (Republic of)

Anupam Yadav
Department of Mathematics
Dr. BR Ambedkar National
Institute of Technology
Jalandhar, India

ISSN 2194-5357   ISSN 2194-5365 (electronic)
Advances in Intelligent Systems and Computing
ISBN 978-3-030-31966-3   ISBN 978-3-030-31967-0 (eBook)
https://doi.org/10.1007/978-3-030-31967-0

This Springer imprint is published by the registered company Springer Nature Switzerland AG
The registered company address is: Gewerbestrasse 11, 6330 Cham, Switzerland

Preface

It is a matter of pride that 5th International Conference on Harmony Search, Soft Computing and Applications (ICHSA 2019) was organized at Kunming, China, during July 20–22, 2019. ICHSA Conference has a glorious history; the earlier editions of the conference were organized at South Korea, Spain and India. This fifth edition of ICHSA were organized in a joint meeting of 15th International Conference on Natural Computation, Fuzzy Systems and Knowledge Discovery (ICNC-FSKD 2019) along with the ICHSA community. The first and the second meetings of the conferences were held at Korea University, South Korea, third meeting of ICHSA was held at Bilbao, Spain, and the fourth meeting was held at BML Munjal University, Gurugram, India. This book is an outcome of the papers presented at ICHSA 2019.

In this book, a collection of good articles have been edited on various topics related to harmony search and soft computing techniques. The balance of swarm intelligence, evolutionary algorithms and their various real-life applications are compiled to provide the consolidated state-of-the-art status of research going in this area. A deep discussion on neural networks is presented for the forecasting, image processing and its other applications. Few articles are presented to address the current challenges in water resource systems and their solutions using soft computing techniques. A similar study is included in seismic events and the earthquake engineering solutions. Some recent advances in harmony search algorithm are added for the researchers working on this. The suitability of soft computing techniques in big data, crowd evacuation strategies, adaptive learning system, economic impact analysis, cyber-attack detection, urban drainage system, water management model, feature selection and inventory systems is presented in this book.

In conclusion, the edited book comprises papers on diverse aspects of soft computing techniques such as nature-inspired optimization algorithms, metaheuristics and their application in areas, such as water resources, urban drainage systems and robotics, pattern recognition, forecasting models and optimization of complex real-life as well as test problems. There are theoretical aspects of the topics mentioned here along with the empirical studies.

July 2019

Joong Hoon Kim
Zong Woo Geem
Donghwi Jung
Do Guen Yoo
Anupam Yadav

Contents

Improved Gray-Neural Network Integrated Forecasting Model Applied in Complex Forecast 1
Geyu Huang, Zhiming Zhang, and Jun Zhang

Design of AGV Positioning Navigation Control System Based on Vision and RFID 16
Jianze Liu, Yan Wang, Jun Sheng, Yangcheng Zhang, Jialin Qi, and LongQi Yu

Improved Image Retrieval Algorithm of GoogLeNet Neural Network 25
Jun Zhang, Yong Li, and Zixian Zeng

A Review of Face Anti-spoofing and Its Applications in China 35
Bizhu Wu, Meng Pan, and Yonghe Zhang

An Event Based Detection of Internal Threat to Information System 44
Zheng Li and Kun Liu

Loop Closure Detection for Visual SLAM Using Simplified Convolution Neural Network 54
Bingbing Xu, Jinfu Yang, Mingai Li, Suishuo Wu, and Yi Shan

Development of Optimal Pump Operation Method for Urban Drainage Systems 63
Yoon Kwon Hwang, Soon Ho Kwon, Eui Hoon Lee, and Joong Hoon Kim

Parameter Estimation of Storm Water Management Model with Sewer Level Data in Urban Watershed 70
Oseong Lim, Young Hwan Choi, Do Guen Yoo, and Joong Hoon Kim

Multiobjective Parameter Calibration of a Hydrological Model Using Harmony Search Algorithm 76
Soon Ho Kwon, Young Hwan Choi, Donghwi Jung, and Joong Hoon Kim

Application of Artificial Neural Network for Cyber-Attack Detection in Water Distribution Systems as Cyber Physical Systems   82
Kyoung Won Min, Young Hwan Choi, Abobakr Khalil Al-Shamiri, and Joong Hoon Kim

Multi-objective Jaya Algorithm for Solving Constrained Multi-objective Optimization Problems   89
Y. Ramu Naidu, A. K. Ojha, and V. Susheela Devi

DSP-Based Implementation of a Real-Time Sound Field Visualization System Using SONAH Algorithm   99
Zhe Zhang, Ming Wu, and Jun Yang

The Economic Impact Analysis of the 1994 Northridge Earthquake by Deep Learning Tools   111
Zhengru Tao, Lu Han, and Kai Bai

SWT-ARMA Modeling of Shenzhen A-Share Highest Composite Stock Price Index   122
Jingyi Wu

The Application of Text Categorization Technology in Adaptive Learning System for Interpretation of Figures   130
Weibo Huang, Zhenpeng He, and Xiaodan Li

A Method for Extracting and Simplifying the Stray Capacitance Matrix of the Dry-Type Smoothing Reactor   139
Tingting Li, Shaoyan Gong, Feng Ji, Chong Gao, and Jianhui Zhou

Coordinated, Efficient and Optimized Crowd Evacuation in Urban Complexes   148
Huabin Feng, Shiyang Qiu, Peng Xu, Shuaifei Song, Wei Zheng, and Hengchang Liu

The Application of Big Data to Improve Pronunciation and Intonation Evaluation in Foreign Language Learning   160
Jin-Wei Dong, Yan-Jun Liao, Xiao-Dan Li, and Wei-bo Huang

Serviceability Assessment Model Based on Pressure Driven Analysis for Water Distribution Networks Under Seismic Events   169
Do Guen Yoo, Chan Wook Lee, Seong Hyun Lim, and Hyeong Suk Kim

Research on an Evaluation of the Work Suitability in the New First-Tier Cities   175
Li Ma, Shi Liu, Longhai Ai, and Yang Sun

Optimal Path Planning in Environments with Static Obstacles by Harmony Search Algorithm   186
Eva Tuba, Ivana Strumberger, Nebojsa Bacanin, and Milan Tuba

Online Feature Selection via Deep Reconstruction Network 194
Johan Holmberg and Ning Xiong

**4-Rule Harmony Search Algorithm for Solving Computationally
Expensive Optimization Test Problems** 202
Ali Sadollah, Joong Hoon Kim, Young Hwan Choi,
and Negar Karamoddin

**An Empirical Case Study for the Application of Agricultural
Product Traceability System** 210
Yu Chuan Liu and Hong Mei Gao

**Study on the Coordinating Policy of Multi-echelon
Inventory System** 218
Tao Yang

**Remote Intelligent Support Architecture for Ground Equipment
Control in Space Launch Sites** 226
Litian Xiao, Mengyuan Li, Kewen Hou, Fei Wang, and Yuliang Li

**Correlation Analysis and Body-Weight Estimation of Yak
Traits in Yushu Prefecture** 235
Xiaofeng Qin, Yu-an Zhang, Meiyun Du, Rende Song, Chenzhang,
and Zijie Sun

**The Macroeconomic Influence of China Futures Market:
A GARCH-MIDAS Approach** 244
Ruobing Liu, Jianhui Yang, and Chuanyang Ruan

Author Index 253

Improved Gray-Neural Network Integrated Forecasting Model Applied in Complex Forecast

Geyu Huang[(✉)], Zhiming Zhang, and Jun Zhang

Engineering University of PAP, Weiyang District, Xi'an 710086, China
huanggy8@mail2.sysu.edu.com

Abstract. Current industry talent demand forecast has a high complexity and non-linearity. The gray method GM(1,1) is suitable to deal with the problem of uncertain forecast with low richness of historical data without consistency, and Back-Propagation Neural Network model (BPNN) is adopted to analyze the influence of current influencing factors. The application of the improved gray-neural network integrated forecasting model in the complex nonlinear forecast is studied in combination with the case of talent demand of a certain city in northern China. Combining the GM(1,1) model, metabolism and background value optimization as the Improved Metabolic GM (1,1) model (IMGM), the forecast result of IMGM is used as the input to train BPNN for improving the forecast accuracy. And the computer simulation flow of talent demand is designed. The result of modeling example shows that the accuracy of improved integrated forecasting model IMGM-BPNN is higher than the conventional model's.

Keywords: IMGM-BPNN · Improved Metabolic GM(1,1) · Talent demand forecast

1 Introduction

Due to the further deepening of China's reform and opening up, various industries in the cities have developed rapidly in recent years. The development mode is greatly different from the previous industrial evolution, and the development pattern is in the leap form. Therefore, the development rule is highly nonlinear and highly complex. In addition, as cities with different levels of development undergo obvious changes over time, the total demand for talents and the structure of talents' academic qualifications have also changed significantly, which has profoundly affected the demand for talents in various industries, making people lack a reference system when making forecasts. Undoubtedly, talent demand forecasting is a complex nonlinear forecasting process [1].

In the process of talent forecast, the complex situation caused by the obvious non-linear characteristics of historical and current influencing factors should be considered. Because of the lack of long-term comprehensive talent history data statistics, considering the discontinuity of data and the lack of long-term comprehensive talent history data statistics, the usual regression method cannot be used to analyze historical data [2].

J. H. Kim et al. (Eds.): ICHSA 2019, AISC 1063, pp. 1–15, 2020.
https://doi.org/10.1007/978-3-030-31967-0_1

2 Related Work

From the perspective of neural network prediction, it has better performance in ana-lyzing non-linear data and time-frequency information. Zecchin et al. [3] develop a short-term glucose prediction algorithm based on a neural network that, in addition to past CGM readings, also exploits information on carbohydrates intakes quantitatively described through a physiological model. Wang et al. [4] take the wavelet packet multilayer perceptron (WP-MLP) as a feature extraction method to obtain time-frequency information. Ren et al. [5] extend the single hidden layer network with random weights to eight variants based on the presence or absence of input layer bias, hidden layer bias and direct input-output connections. It has the shortest training time compared with the other reported methods. Zhu and Wang [6] propose an intelligent trading system using support vector regression optimized by genetic algorithms and multilayer perceptron optimized with genetic algorithms. This system can well process the highly noisy and non-linear data but doesn't consider that data might be in a low richness.

From the perspective of talent demand forecasting methods, the main methods adopted at present can be divided into qualitative and quantitative methods. The former includes Delphi method, descriptive method and empirical prediction method. The latter includes regression model, production function model and grey model. Among them, grey model is one of the main methods for talent demand forecasting. For example, Song [7] and Yang [8] use grey system theory to build GM(1,1) model for total talent demand forecasting of new industries. Min [9] constructs the grey prediction model of talent demand and makes a comparative analysis with the application results based on the time series and the main factor GDP regression model. Wang [10] combines the time-varying growth rate model, linear regression model and gray GM (1,1) prediction model for prediction. However, the above talent demand prediction based on the gray model only adopts a single conventional GM (1,1) model without any improvement. Some models have the problem of too long span of years in the time selection of sample data, which leads to the distortion of the model and the low accuracy of the proposed measurement.

Based on GM(1,1) basic model and metabolic GM(1,1) model, with full consid-eration of the influence of new information such as disturbance factors and driving factors, and by means of MATLAB software programming calculation and test, Wang et al. [11] predicted the aged population aged 65 and above in Chongqing from 2015 to 2020. This model has the top level of accuracy and is suitable for medium and long term prediction. However, metabolism doesn't take into account the expansion of the model's scope of adaptation, while it improves the smoothness of the data. Liu et al. [12] reconstructed an accurate and optimized background value construction formula on the basis of the existing background value construction method, which has high fitting and prediction accuracy, and is applicable to the modeling of high and low exponential growth sequences at the same time. Therefore, both metabolic GM (1,1) model and background value optimization can improve the accuracy of model forecast.

The gray method is just suitable to deal with the problem of uncertain forecast with low richness of historical data without consistency, so the gray dynamic model GM (1,1) is adopted to analyze the influence of historical factors on future talent demand in this paper. And based on the metabolic mechanism and background value optimization to improve it. It is necessary to consider not only the role of historical influencing factors, but also the current influencing factors such as development planning, industry current development trend, industry growth limit, maximum output value constraint and industry average labor productivity. However, these current influencing factors have strong complexity and non-linearity. The general correlation analysis method has a very low precision when applied to the complex multi-factor correlation analysis. What's more, the correlation analysis model cannot deal with the nonlinear correlation problems. Therefore, an advanced intelligent algorithm model Back-Propagation Neural Network model (BPNN) is adopted to analyze the influence of current influencing factors on future talent demand. BP has the advantages of high forecast accuracy and good performance in dealing with complex nonlinear problems, suitable for this kind of current impact factor analysis. The application of the improved gray-neural network integrated forecasting model IMGM-BPNN (Improved metabolic GM (1, 1) model and Back-Propagation Neural Network model) in the complex nonlinear forecast is studied in combination with the case of talent demand of a certain city in northern China below.

3 Improved Integrated Forecasting Model IMGM-BPNN

3.1 Improved Metabolic GM(1,1) Forecasting Models

The Grey forecasting method establishes a GM model extended from the past to the future based on the known or unknown information from the past and the present, to identify the system's development trend, providing the basis for planning and decision-making. The Grey forecasting model makes projections of the number size for time series, in which the randomness is weakened while the certainty is enhanced. At this point, solving on the generated level to get the generating function, the sequence forecast needed is established whose forecasting model is a differential equation of first order only including one variable, denoted as GM (1,1) model.

GM(1,1) model is the core of gray forecast [13]. It is a first-order differential equation model for single variable forecast, and its discrete time response function is approximately exponential. The GM(1,1) model is established by:

The original non-negative time series is $X^{(0)} = \{X^{(0)}(1), X^{(0)}(2), \ldots, X^{(0)}(n)\}$, the accumulating generation function is $X^{(1)}(t)$ [14], thus

$$X^{(1)}(t) = \sum_{m=1}^{i} X^{(0)}(m), t = 1, 2, \cdots, n \tag{1}$$

The albinism differential equation of GM(1,1) is

$$\frac{dX^{(1)}}{dt} + aX^{(1)} = u \tag{2}$$

a is the parameter to be identified in (2), which can also be denoted as the evolution parameter. u is the endogenous variable or gray action. Set the identified vector $\hat{a} = \begin{pmatrix} a \\ u \end{pmatrix}$, apply the least square method on $\hat{a} = (B^T B)^{-1} B^T y$ and get

$$B = \begin{vmatrix} -\frac{1}{2}\left(X^{(1)}(1) + X^{(1)}(2)\right) & 1 \\ -\frac{1}{2}\left(X^{(1)}(2) + X^{(1)}(3)\right) & 1 \\ \cdots\cdots & \cdots \\ -\frac{1}{2}\left(X^{(1)}(n-1) + X^{(1)}(n)\right) & 1 \end{vmatrix}$$

$$y = \begin{vmatrix} X^{(0)}(2) \\ X^{(0)}(3) \\ \cdots \\ X^{(0)}(n) \end{vmatrix} \tag{3}$$

Thus, the discrete time response function of Grey forecast is

$$X^{(1)}(t+1) = \left(X^{(0)}(1) - \frac{u}{a}\right)e^{-at} + \frac{u}{a} \tag{4}$$

$X^{(1)}(t+1)$ is the accumulating forecasting value which is acquired, then restore this forecasting value and get

$$\hat{X}^{(0)}(t+1) = \hat{X}^{(1)}(t+1) - \hat{X}^{(1)}(t), (t = 1, 2, 3, \cdots n) \tag{5}$$

In the past, the talent demand forecast based on Gray model was basically based on the conventional GM (1,1) model, and some interference factors in the development process of the system were ignored, thus affecting the forecast accuracy, especially the accuracy in the medium and long term. The metabolic GM (1,1) model can well compensate for this deficiency. The principle is to add the latest data $\hat{X}^{(0)}(t+1)$ forecast by the GM (1,1) model to the original data sequence $X^{(0)}$, and then remove the oldest data $X^{(0)}(1)$ to ensure the invariant dimensions of the data sequence. Then the general GM (1,1) model was established repeatedly with the latest data sequence until the forecast target is completed, that is, the GM(1,1) models based on metabolism.

The metabolic GM (1,1) model above is a very effective optimization compared with the conventional GM (1,1) model. It can significantly improve the forecast accuracy. However, metabolism doesn't optimize the background values, so there are still defects. Wei and Yuxiang [15] pointed out that the error in the trapezoidal formula construction method between

$$\frac{1}{2}\left(x^{(1)}(k) - x^{(1)}(k-1)\right) \tag{6}$$

and

$$\int_{k-1}^{k} x^{(1)}(t)dt, k = 2, 3, \cdots, n \tag{7}$$

is the main reason for the low accuracy of GM (1,1) model, and proposed a more accurate background value construction formula:

$$Z^{(1)}(k+1) = \frac{x^{(1)}(k+1) - x^{(1)}(k)}{\ln(x^{(1)}(k+1) - x^{(1)}(1)) - \ln(x^{(1)}(k))} - \frac{x^{(1)}(1)x^{(1)}(k)}{x^{(0)}(k+1) - x^{(1)}(1)}, \tag{8}$$
$$k = 1, 2, \cdots, n-1$$

By optimizing the background value, the accuracy of model fitting and forecast is improved. In general, both metabolic GM (1,1) model and background value optimization can improve the accuracy of model forecast, but we combine these two methods as the Improved Metabolic GM (1,1) model, and then use the forecast result of IMGM as the input to train BP Neural Network for improving the forecast accuracy. In order to test the forecast model IMGM-BPNN, it was formed and compared with the conventional GM (1,1) model, metabolic GM (1,1) model and background value optimization GM (1,1) model among fitting accuracy. According to this model, the talent demand of a certain city can be forecast. In order to compare the forecast accuracy of the four models, the mean absolute percent error (MAPE) was used to represent the forecast error of each model (Fig. 1).

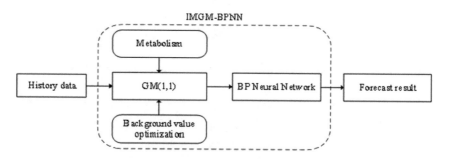

Fig. 1. Modeling process of IMGM-BPNN

3.2   Back-Propagation Neural Network

BPNN is a multi-layer network formed by the interconnection of input layer, output layer and one or more hidden layer nodes. This structure enables the multi-layer feedforward network to establish an appropriate linear or nonlinear relationship between input and output without restricting the network output between −1 and 1 [16] (Fig. 2).

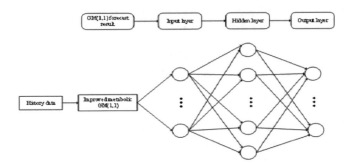

Fig. 2. BP network structure in IMGM-BPNN

BP algorithm obtains the appropriate linear or nonlinear relationship between input and output through the "training" event. The process of "training" can be divided into two stages: forward propagation and backward propagation [17].

Forward propagation:

Step1. Take a sample P_i, Q_j from the sample set, pass P_i to the network;

Step2. Calculate the error measure E_i and actual output

$$O_t = F_L\left(\cdots\left(F_2\left(F_1(P_tW^{(1)})W^{(2)}\right)\cdots\right)W^{(L)}\right) \tag{9}$$

Step3. Repeat adjusting the weights $W^{(1)}, W^{(2)}, \ldots, W^L$ until $\sum E_i < \varepsilon$

Backward propagation (Error propagation):

Step1. Calculate the difference between actual output O_p and ideal output Q_i;

Step2. Use this difference $E_i = \frac{1}{2}\sum_{j=1}^{m}\left(Q_{ij} - O_{ij}\right)^2$, that is, the error of the output layer to adjust the weight matrix of the output layer;

Step3. User the error $E_i = \frac{1}{2}\sum_{j=1}^{m}\left(Q_{ij} - O_{ij}\right)^2$ to estimate the error of the direct leading layer in the output layer, and then estimate all of the other error of the previous layers;

Step4. Use these estimate to modify the weight matrix, forming a process of transmitting the errors shown at the output end to the input end step by step in the direction opposite to the output signal.

Network error measure for the entire sample set:

$$E = \sum_i E_t \tag{10}$$

The function of the neural network is S(Sigmoid) function:

$$f(x) = \frac{1}{1 + \exp(-x + \theta)} \tag{11}$$

But, the sigmoid tangent function tansig is adopted instead:

$$f(x) = \frac{2}{1 + \exp(-2x + \theta)} - 1 \tag{12}$$

θ is the Error criteria adjust weights in the thresholding algorithm to ensure the deviation $E = \frac{1}{2}\sum_{pl}(D_{pl} - Y_{pl})^2 = \frac{1}{2}\sum_{pl}\varepsilon_i^2$ is smallest. D_{pl}, Y_{pl} are the expected output value and the actual output value of the output unit in the input mode P respectively.

3.3 IMGM-BPNN Applied in Talent Demand Forecast

In the global forecast process, the two models above are combined as the improved gray-neural network integrated forecast model IMGM-BPNN to forecast the talent demand of a certain city in the future. The steps of model application are shown below:

Step1. Input raw data;
Step2. Apply the Improved metabolic gray model to forecast, and obtain the forecast sequence;
Step3. Take the forecast value as the input and the original data as the expected value, to train the BP Neural Network obtaining the corresponding weights and thresholds;
Step4. Input the month, year or other elements that need to be forecast, and get the forecast value with considerable accuracy.

According to these steps, the flow chart of computer simulation algorithm for talent demand forecast is given (Fig. 3).

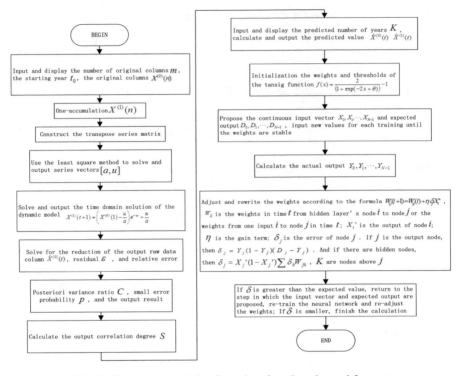

Fig. 3. Computer simulation flow chart for talent demand forecast

4 Talent Demand Analysis of a Certain City

4.1 Talent Aggregate Demand Analysis

Figure 4 shows the statistics of the monthly aggregate demand of talents from a certain city in northern China (named A-city below). Figure 5 analyzes the employment situation of A-city by comparing the aggregate demand for talents between 2011–2012 and 2016–2017.

The employment situation of A-City is analyzed as follows through Figs. 4 and 5.

- In each year, February to April is the big peak of A-City's recruitment, while August to October is the small peak of a-city's recruitment. During this period, there are many employment opportunities, and a large job gap appears, so job seekers should seize the opportunities.
- A-city's annual demand for talents in March is much higher than that in other months, so job seekers should mainly focus on March for job hunting and preparation.
- When comparing the talent demand year by year, especially the comparison between 2011–2012 and 2016–2017, it is found that the talent demand of A-City decreases year by year.

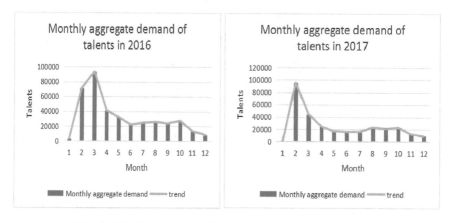

Fig. 4. Monthly aggregate demand of talents in 2016 and 2017

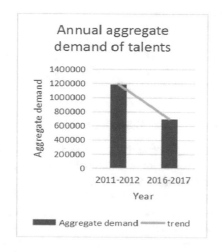

Fig. 5. Annual aggregate demand of talents

4.2 Talent Structure Analysis

The academic structure of A-City's annual talent demand is shown in Table 1. According to the analysis in Table 1, overall, the structure of educational background required by talents is basically stable with no major changes in each year. At the same time, the proportion of the posts with bachelor's degree in the total number of A-City talents demand is very small, indicating that the bachelor's degree is not an ideal education background for A-City. Graduates or job seekers, either enter technical secondary school or junior college to learn professional skills and graduate earlier, or continue to study for a master's degree and doctor's degree, both of which are more popular with recruiters.

Table 1. Annual academic structure of talent demand

Year	2011–2012	2015	2016	2017	2018
Below bachelor degree	54%	59%	58%	59%	61%
Bachelor degree	3%	4%	7%	8%	4%
Above bachelor degree	43%	37%	35%	33%	35%

The following is the analysis of the academic structure of the four sectors with the largest number of talents in A-City in recent years. Table below shows the academic structure of talents in the four sectors in 2018.

Table 2. Academic structure of talents in four sectors in 2018

Academic degree	Below bachelor degree	Bachelor degree	Above bachelor degree
Wholesale and retail trade	59%	4%	37%
Scientific research and technical service	51%	1%	48%
Leasing and business service	70%	8%	22%
Others	61%	4%	35%

It can be obtained by analyzing Tables 1 and 2,

- Wholesale and retail trade, Leasing and business service and others require less than a bachelor's degree account for a prominent majority, and these three sector directions are suitable for applicants with less than a high degree, and at the same time, a certain number of people with a high degree are also recruited.
- In Scientific research and technical service, the number of posts requiring bachelor's degree or above is basically flat. And comparing with the other three sector directions, this sector is inclined to recruit people with high academic level.

5   Forecast Results

When training the neural network in MATLAB R2017a, a three-layer network topology is adopted, and the number of hidden layer neurons is 10. Other parameters of BPNN are set as follows (Table 3):

Table 3. Parameter setting of BPNN

Parameter	Setting
Minimum training rate	0.2
Target error	$1e-6$
Maximum iteration number	10000
Hidden layer transfer function	tansig
Output layer transfer function	purelin
Network training function	traingd

5.1 Forecast Results of Talent Aggregate Demand

After entering the historical data of the total number of talents in a certain city and the current influence factor data into the forecast simulation program, the specific forecast results of A-City's total talent demand are shown in Table 4. It can be seen from the above data that A-City's talent demand started to rise after a sharp decline in 2018, and it is expected that by the end of the 13th five-year plan, its total talent demand will reach 269,347 people, and the annual growth rate of total talent demand will be about 5% from 2019 to 2021, with the increasing of employment opportunities.

Table 4. Forecast aggregate demand of talents

Year	Aggregate demand of talents
2018	90600
2019	245738
2020	258103
2021	269347

5.2 Forecast Results of Sector Structure

The improved IMGM-BPNN integrated model simulation program is run again according to the previous total forecast results and structural analysis data. In recent years, the concrete forecast results of the talent demand structure in A-City's top four sector concrete forecast results are as shown in Table 5. It can be seen from the above data, talent demand of A-City's Scientific research and technical service has been rising steadily, and demand of wholesale and retail trade continues to fall. A-City needs more and more technical talent, and continue to shift from labor-intensive city to technological innovation-oriented one.

Table 5. Forecast sector structure

Sector	Year		
	2019	2020	2021
Scientific research and technical service	15%	17%	18%
Leasing and business service	6%	7%	10%
Wholesale and retail trade	33%	30%	25%
Others	14%	14%	15%

6   Precision Analysis

6.1   Comparison of Fitting Accuracy of Different GM(1,1) Models

We fit the job demand data on September 2015 to August 2018 from month to month through 4 GM(1,1) models, lastly 2016 and 2017's total demand is sum by forecast 2016 and 2017's months total sum to calculate the four kinds of the relative error of GM (1, 1) model. On the basis of the comparison of MAPE of the four models, IMGM-BPNN with the least fitting error is selected to forecast the talent demand of A-City in from 2018 to 2021. The fitting data and precision pairs of the four models are shown in the Table 6.

Table 6. Comparison of fitting accuracy of 4 kinds of GM(1,1) models

Year	Actual demand	Normal		Metabolic		Optimized background value		IMGM-BPNN	
		Forecast	Relative error (%)	Forecast	Relative error (%)	Forecast	Relative error (%)	Forecast	Relative error (%)
2016	391628	395153	0.9	393194	0.4	395150	0.9	391824	0.05
2017	306567	304421	0.7	305647	0.3	304399	0.7	395149	0.4
	MAPE		0.8		0.35		0.8		0.225

6.2   Precision Analysis of Talent Aggregate Demand Forecast

According to the actual total number of talents in this city and the forecast value of Table 7 in 2015, and residual error formula $e(k) = x(k) - \hat{x}(k)$ ($x(k)$ is raw data, $\hat{x}(k)$ is forecast value), the residual error is calculated as shown in the table. Residual mean formula: $\bar{e} = \frac{1}{n} e_{total} = \frac{1}{n} \sum_{k=1}^{n} e(k)$, apply it to calculate and get $\bar{e} = 0.0575$. The variance of talent aggregate demand's actual value is $s_1 = \sqrt{\frac{1}{n} \sum_{k=1}^{n} (x(k) - \bar{x})^2}$, $\bar{x}$ is the mean of actual data $\bar{x} = \frac{1}{n} \sum_{k=1}^{n} x(k)$. The variance of residual error is $s_2 = \left\{ \frac{1}{n} \sum_{k=1}^{n} [e_c(k)]^2 \right\}^{\frac{1}{2}}$, and $e_c(k) = |e(k) - \bar{e}|$. It can be calculate that $s_1 = 3.98$, $s_2 = 0.33$, so the posterior difference ratio is $c = s_2/s_1 = 0.083$.

Table 7. Precision analysis of talent aggregate demand forecast

Year	2011	2012	2013	2014	Total
Forecast value	59.48	54.33	51.21	49.61	214.63
Actual value	60.01	54.11	51.1	49.64	214.86
$e(k)$		0.53	−0.22	−0.11	0.03	0.23
$s_1(k)$		6.29	0.4	2.62	4.08	3.98
$e_c(k)$		0.3	0.45	0.34	0.2	1.29
$s_2(k)$		0.09	0.20	0.12	0.04	0.33

If $e_c(k) < \alpha s_1$, the error of forecast value $\hat{x}(k)$ is regard to be small. Usually in the error test of talent demand forecast, set $\alpha = 0.675$, then $\alpha s_1 = 2.69$. All $e_c(k)$ are less than 2.69, which means all $\hat{x}(k)$ are acceptable, then small error probability $P = P\{e_c(k) < \alpha s_1\} = 1$.

In the precision analysis process, when the posterior difference ratio $c < \varepsilon$ and the small error probability $P > \varphi$, the model can be considered credible, the forecast accuracy is good, and the forecast value is acceptable. Here, the indicators are selected strictly according to the parameters in the field of talent demand forecast, $\varepsilon = 0.35$, $\varphi = 0.95$ (first-order accuracy threshold), then $c = 0.083 < 0.35$, $P = 1 > 0.95$. It can be concluded that when apply GM-BPNNM integrated model to forecast talent aggregate demand, the model has high reliability, the forecast accuracy reaches the highest level and the forecast value is available.

6.3   Precision Analysis of Sector Structure Forecast

The method as same as that analyze the accuracy of the previous total forecast results is used, take the same parameters $\alpha = 0.675$, $\varepsilon = 0.35$, $\varphi = 0.95$ and get $c = 0.076 < 0.35$, $P = 1 > 0.95$. Therefore, the sector structure forecast is credible, the accuracy reaches one level, and the forecast value is available.

To sum up, the integrated model is successful in forecasting talent demand.

7   Conclusion

GM(1,1) model is a nonlinear model with small sample size and high forecast accuracy, which is very suitable for the forecast of small sample and poor data. However, the conventional GM(1,1) model has the defects of initial sequence and background value setting. Therefore, the improved metabolic GM(1,1) model, background value optimization model and BP Neural Network are integrated to improve the conventional GM(1,1) model, taking a certain city in northern China as an example to make an talent demand analysis. The result of modeling example shows that the accuracy of improved integrated forecasting model IMGM-BPNN higher than the conventional model, and it is superior to the other two models having obvious forecast accuracy advantages.

According to the comparison of the forecast results, by the end of the 13th five-year plan, the demand for talent in this northern city will reach 269,347, with an average annual growth rate of nearly 5%. At the same time, more and more talents are needed in scientific research and technical service industries, and the requirements for academic qualifications are higher and higher. In the future, in terms of talent increment and talent structure, the city should continue to explore how to formulate more scientific and effective talent introduction strategies, further release talent vitality, and realize the transformation from a big city to a strong one with talents.

References

1. Xia, J., Xu, H.: Forecasting tourism income of Yunnan province by gray neural network. Value Eng. **37**(21), 104–108 (2018)
2. Yuan, P.-w, Song, S.-x, Dong, X.-q: Study on fire accident prediction based on optimized grey neural network combination model. J. Saf. Sci. Technol. **10**(3), 119–124 (2014)
3. Zecchin, C., Facchinetti, A., Sparacino, G., De Nicolao, G., Cobelli, C.: A new neural network approach for short-term glucose prediction using continuous glucose monitoring time-series and meal information. In: 2011 Annual International Conference of the IEEE Engineering in Medicine and Biology Society, pp. 5653–5656 (2011)
4. Wang, L.P., Teo, K.K., Lin, Z.P.: Predicting time series with wavelet packet neural networks. In: 2001 IEEE International Joint Conference on Neural Networks (IJCNN 2001), pp. 1593–1597 (2001)
5. Ren, Y., Suganthan, P.N., Srikanth, N., Amaratunga, G.: Random vector functional link network for short-term electricity load demand forecasting. Inf. Sci. **367–368**, 1078–1093 (2016)
6. Zhu, M., Wang, L.P.: Intelligent trading using support vector regression and multilayer perceptrons optimized with genetic algorithms. In: 2010 International Joint Conference on Neural Networks (IJCNN 2010) (2010)
7. Song, D.: Analysis on the demand and training mode of professionals in strategic emerging industries – taking Jiangsu province as an example. Reform Opening up **9**, 56–59 (2015)
8. Yang, L., He, Z., Han, F.: Strategic emerging industry talent demand forecast and countermeasures for Hunan province. Forum Sci. Technol. China **1**(11), 85–91 (2013)
9. Min, X.: Talent demand analysis based on gray prediction model GM(1,1). Sci. Technol. Manag. Res. **6**, 25 (2005)
10. Wang, C.: Construction of talent demand forecasting model for strategic emerging industries – taking Jiangsu province as an example. Sci. Technol. Assoc. Forum (second half), **2**, 141–142 (2012)
11. Wang, N., Zhang, S., Zeng, Q.: Research on aging prediction of Chongqing population based on metabolic GM(1,1) model. Northwest Popul. J. **38**(1), 66–70 (2017)
12. Liu, L., Wang, H., Wang, B.: GM(1,1) model optimization based on background value construction method. Stat. Decis. **277**(1), 153–155 (2009)
13. Kong, X., Wang, L., Feng, Y.-h.: Review and prospects of the application status of GM (1,1). J. Qilu Univ. Technol. **32**(6), 49–53 (2018)

14. Wang, H.: Research on gray neural network prediction model. North China University of Water Resources and Electric Power

15. Wei, Y., Yuxiang, L.: Manpower demand forecasting of strategic emerging industry in china: based on grey system methodology. In: Portland International Conference on Management of Engineering & Technology. IEEE (2015)

16. O'Brien-Pallas, L., Baumann, A., Donner, G., et al.: Forecasting models for human resources in health care. J. Adv. Nurs. **33**(1), 10 (2001)

17. Goyal, V., Deolia, V.K., Sharma, T.N.: Robust sliding mode control for nonlinear discrete-time delayed systems based on neural network. Intell. Control Autom. **1**, 75–83 (2015)

Design of AGV Positioning Navigation Control System Based on Vision and RFID

Jianze Liu, Yan Wang$^{(\boxtimes)}$, Jun Sheng, Yangcheng Zhang, Jialin Qi,
and LongQi Yu

Beijing SpaceCraft, Youyistr. 104, Beijing 100094, China
345929552@qq.com

Abstract. A design of composite navigation control system based on RFID and vision is proposed for problems, such as low positioning, poor stability, high cost, of the present AGV navigation ways. AGV recognizes the station with RFID firstly, then AGV builds an image recognition system with visual technology to achieve the accurate positioning. In this way, AGV can accomplish complex navigation tasks more accurately and efficiently. Low recognition precision of guides and noise interference can be solved with software by using gray image segmentation, image edge extraction, image denoising and linear fitting. The positioning accuracy of AGV is raised to 5 mm, meanwhile the angle precision is raised to 0.1° in application. The system not only satisfies the continuous positioning in large space, but also meets the requirement of high precision positioning, and realizes industrial automation.

Keywords: AGV · RFID · Vision · Composite navigation

1 Preface

Automated guided vehicles (AGV) is an important branch of intelligent logistics. AGV, which known as autonomous unmanned vehicles, is designed for industrial applications [1, 2]. It can intelligently walk and stop at a designated station according to certain path planning and operation requirements under the control of the dispatching system [3], then complete a series of further tasks such as transporting goods, charging, clamping and so on. Intelligent manufacturing has listed as the goal of enterprise manufacturing and logistics transfer systems by commercial, marine, and military industry with the introduction of the industrial 4.0 strategic concept, and AGV automatic transfer has become the only way to realize industrial automation.

1.1 System Overview

In this case, the wheel structure of AGV are composed of four Mecanum wheels, which can realize all-round motion, including straight line, horizontal line, oblique line, arbitrary curve movement, zero radius of gyration, etc. [4, 5]. The structure effectively reduces the requirements of the AGV for working area and expands the working range of the AGV.

© Springer Nature Switzerland AG 2020
J. H. Kim et al. (Eds.): ICHSA 2019, AISC 1063, pp. 16–24, 2020.
https://doi.org/10.1007/978-3-030-31967-0_2

The AGV control system consists of three controllers: a motion controller, a navigation controller and a lift controller (see Fig. 1). The motion controller mainly control the wheeling movement, in the meantime, the obstacle information collected by laser obstacle avoidance sensor to guarantee the AGV navigation controller have normal navigation, automatic charging and the wireless communication function when doing reliable movement. Through four CCD cameras and RFID, AGV can achieve automatic cruising and positioning function. AGV can receive dispatch system instruction and feedback status information through the wireless communication function module of AGV and dispatch system. When AGV reaches the designated station, the lift controller controls the motor to lift or drop the product to the specified height and complete the work.

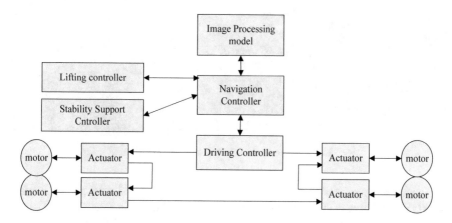

Fig. 1. AGV system structure

1.2   Hardware Design of Navigation Control System

Communication mode, communication speed, system accuracy, AGV working mode and functional requirements should be considered comprehensively when designing the hardware of the control system.

Since LPC1788 as the core of ARM control chip, the navigation controller is designed by its system. The chip is an ARM Cortex-M3 microcontroller with integrated LCD image controller, which includes LCD controller, CAN bus controller, 512 KB on-chip flash memory, 96 KB SRAM, 4 KB on-chip EEPROM, 4 UART, 12 Bit ADC and 10-bit DAC. The navigation controller includes a CAN bus drive circuit, a UART circuit drive and its switching circuit, an IO signal processing circuit, a signal level conversion circuit, and a power supply circuit.

The AGV connects seven nodes via the CAN bus, which include four CCD cameras, a lift controller, a motion controller, and a laser obstacle avoidance sensor. The RFID module and the ZIGBEE wireless module communicate with the navigation controller through RS232 communication mode.

2 Navigation Control Strategy

2.1 Method of Visual Positioning Precisely

The AGV can achieve automatic cruising function by using image recognition processing technology, which need to lay a yellow guide tape or paint a yellow ribbon as a guiding path. Figure 2 shows the straight, cross and parking signs. The way that guide line preset will define the travel path of the AGV, which can greatly reduce the range of AGV activity and improve the space utilization and transfer efficiency of the workshop [7, 8].

Fig. 2. The straight, cross and parking signs

When cruising, install a vision system on the front, rear, left side, and right side of the AGV: including camera, light source, lens, and position adjustment mechanism. To complete the identification of the guide line, the transformation matrix of image coordinate system and vehicle coordinate system should be established through the guide line image recognition processing, which include average gray level, binarized segmentation, feature extraction, constraint conditions and denoising processing, line equation fitting. [9–12] To achieve automatic transfer of AGV, the navigation controller will plans the AGV travel direction and angle information according to the relevant information of the current sensor feedback, then release a running order to AGV motion controller, finally, the motion controller can match the speed and direction of each motor reasonably.

2.2 RFID Positioning Method

RFID devices obtain information by using RFID tags and readers [6]. Laying a navigation belt at the operating site of the AGV to layout cruise route. An RFID card is attached at the crossroads and the station, each RFID card has different information, corresponding to different positions (see Fig. 3). Because the RF card has limited recognition range, to prevent the problem that the station information cannot be accurately identified in the dense station, the RF card reader with a relatively small recognition range (about 1 m) is selected, and the station spacing is required to be more than 2 m.

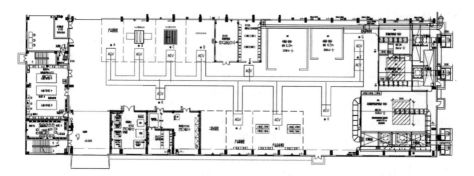

Fig. 3. RFID working condition schematic diagram

After receiving the superior processing instruction, the dispatching system plans a working path of each AGV based on the guiding map and sends it to the AGV. The AGV recognizes the guide line through the on-board CCD and travels in a large operation range; when running to the station, the RFID card reader reads the RF card information to identify the current position, and achieves a rough stop; then the vehicle CCD further Identify the stop sign to achieve precise positioning of the AGV. During the whole operation process, the AGV will feedback to the scheduling system itself in real time, and the scheduling system will make the next path adjustment according to the feedback.

3   Image Processing Algorithm

3.1   Automatic Exposure Algorithm Design of Visual Image Acquisition

Several monitoring square are set in the image, whose size is a × a, and the side length a (unit: pixel) can be selected, the sub-images of several square are distributed as evenly as possible in the entire image. The average gray value of several sub-images is:

$$\bar{G} = (N \cdot a^2)^{-1} \sum_{k=1}^{N} \sum_{i=1}^{a} \sum_{j=1}^{a} g_{ij} \tag{1}$$

The image target brightness is given G0, the scale factor between the current image brightness and the target brightness is:

$$d = G_0/(\bar{G}+1) \tag{2}$$

To ensure the camera does not produce large brightness changes in the light and dark areas, adjust the exposure of the camera in the loop, for each frame of image captured, modify the brightness of the image in real time, which will establish the basis for the stability of the subsequent image processing algorithms [13–15].

3.2   Algorithm Design of Visual Longitudinal Guide Line

The detection of the longitudinal guide line is extremely important in the entire algorithm system. It is not only provided to the AGV control system in the form of mathematical information, but also a prerequisite for the detection of the lateral guide line and the parking sign. The primary task of the longitudinal guideline detection algorithm is to extract the guide lines from the background. In this paper, the contour extraction method is adopted, which is less affected by the change of image brightness and has better robustness to the change of spatial and dark areas.

3.3   Extraction Algorithm Target Contour Pixel

The gray value at (x, y) in an image is represented by f(x, y). The first order differential is:

$$f'(x, y) = \frac{\partial f}{\partial x} + \frac{\partial f}{\partial y} \tag{3}$$

The discretized formula is as follow:

$$\Delta f(i,j) = [f(i+1,j) - f(i,j)] + [f(i,j+1) - f(i,j)] \tag{4}$$

Based on the principle of the greatly change of the target edge gradient, the edge pixels of the target are detected. When Gx and Gy are used to represent the gradient changes in the X and Y directions, the total gradient is as follows:

$$G = \sqrt{G_x + G_y} \tag{5}$$

3.4   Edge Contour Detection

The image is quantized by gradient, and the gradient graph shown in Fig. 4 is obtained. The intensity of edge is stronger than that of non-edge. Based on this principle, an edge contour detection algorithm is designed.

Min(j) is used to represent the minimum gradient of line j in the gradient image. Calculate the average value of the smallest value in the whole image, M represents the total number of rows, then:

$$Min\_Avg = \frac{1}{N} \sum_{j=1}^{M} Min(j) \tag{6}$$

Given an additional threshold T0, the edge pixel detection operator is as follows:

$$\begin{cases} G_{ij} > Min\_Avg + T_0, \text{Edge pixels} \\ G_{ij} \leq Min\_Avg + T_0, \text{Background pixels} \end{cases} \tag{7}$$

Fig. 4. The gradient graph

3.5   Recognition and Classification of Target Pixel

The edge image in Fig. 5 includes pixel of target object, such as edge pixels of guide line, edge pixels of the parking sign, etc., but may also include a stained edge or a non-target edge in the image, and a corresponding algorithm needs to be established to identify and classify the actual target, except the non-actual target pixel.

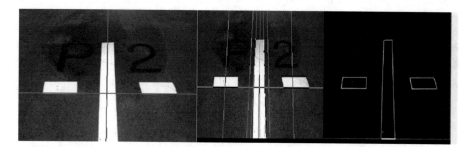

Fig. 5. Image recognition effect

The edge pixel image should be encoded first, because the pixels in the image are arranged in columns, and the pixels are connected in the same target, the different targets are not connected. For a row of pixels, consecutively edge pixels arranged as a unit, denoted as U_j, i, and j represents a row, and i represents the i-th unit of the j-th column. Segmentation of objects in an image using connectivity, a connected area is a possible object, the specific judgment conditions are as follows: the current U_j, i is compared with the unit U_{j+1}, k of the next line, and each unit is traversed. If the two units are connected or have overlapping parts, that is a units in a region, mark them to obtain all the connected areas in the image like above.

For each connected area, count its area, regional horizontal minimum, maximum coordinate, vertical minimum and maximum coordinate. If the stain area of image is small, it can be filtered initially by setting a threshold. If the area of the target area is Area(k), the stain area is Area(k) < T1.

The image may include edge pixels of longitudinal guide line, edge pixels of lateral guide line, parking mark areas or other non-useful target areas after the above filtering. Propose the criteria for identification classifier based on the characteristics of each region are analyzed.

Using the horizontal scanning search strategy, if the jth row is searched, start 0 column → N column search, the first edge pixel position encountered is xi; Start N column → 0 column search, the first edge pixel position encountered is xk, and if the two lateral coordinates satisfy the following conditions, the edge pixels of the longitudinal guide line are initially identified:

$$
\begin{cases}
x_i < x_k \\
|x_i - x_k| > W_1 \,\&\, |x_i - x_k| < W_2
\end{cases}
\tag{8}
$$

Among them, W1 and W2 are two width thresholds. Formula (8) can be used to get the coordinates of the edge pixels of the longitudinal guide, but these pixels do not belong to the edge of the longitudinal guide, so it is necessary to detect the points on the straight line automatically by using the line features. Two sets of edge pixels of the longitudinal guide are represented as: $\{x_{lk}, y_{lk}\}$ and $\{x_{rk}, y_{rk}\}$. The two sets are divided into two parts, each part is divided into the upper part and the lower part: $\{x_{li}^t, y_{li}^t\}$, $\{x_{lm}^b, y_{lm}^b\}$ and $\{x_{ru}^t, y_{ru}^t\}$, $\{x_{rv}^b, y_{rv}^b\}$, that i = 1, 2,......, N1, m = 1, 2,......, N2, u = 1, 2, ..., N3, v = 1, 2,......, N4.

Arbitrary selection of two points: $\{x_{li}^t, y_{li}^t\}$, $\{x_{lm}^b, y_{lm}^b\}$. Obtain the linear equations: $y_{im} = k_{im}x_{im} + b_{im}$ (k_{im}: slope, b_{im}: intercept). The distance from the $\{x_{lk}, y_{lk}\}$ set to the straight line is as follows:

$$
D_{im} = |y_{lk} - k_{im}x_{lk} - b_{im}| / \sqrt{1 + (k_{im})^2}
\tag{9}
$$

If $\{D_m < T_5\}$, the pixel belongs to the line, the number of pixels owned by the line Num_{im} is recorded by statistics. Find all the straight lines to $\{x_{li}^t, y_{li}^t\}$, $\{x_{lm}^b, y_{lm}^b\}$, and do the similar statistics, a *set* of pixel points Num_j are obtained, j = Min(i, m), solve the equation $Num_{Max} = Max(Num_j)$, if $Num_{Max} = Max(Num_j)$ and $Num_{Max} > T_6$, then the corresponding linear equation is the left edge straight line of the longitudinal guide line. A similar method is also used to calculate the right edge pixel, and the linear equation corresponding to the left and right edge lines is obtained respectively: $y = k_l x + b_l$, $y = k_r x + b_r$, and the two lines are further derivative to find the guide line equation, and the offset distance between the camera and the vehicle body coordinate system is obtained. According to the offset distance between the camera and the car body coordinate system, and the field of view range, the relationship between the car body and the current path is derived, thus the final guiding equation is derived.

4 Application and Verification

Image recognition is rich in information and have high precision. which is suitable for indoor environments such as workshops, especially the use environment where the navigation path is complex and the positioning accuracy is high [16]. At present, the AGV that based on RFID and vision guidance has been applied in the national power inspection industry and other different automated transfer workshop. In practical application, the AGV guide line width is set to 10 mm, and the maximum speed is set to 0.7 m/s. The alignment and positioning accuracy can reach 5 mm, and the angular precision is 0.1°. The AGV based on RFID and visual composite navigation has expanded multi-function task modules such as lifting, forklift and clamping according to different operation requirements. At the same time, the multi-vehicle collaborative operation can be completed by planning the path and publishing tasks through the scheduling system (see Fig. 6).

Fig. 6. Image recognition effect

5 Conclusion

Based on the 13th Five-Year Intelligent Equipment development plan, and in order to develop modularization, automation, intelligence and informationization to adapt future production needs, in this paper, a number of advanced technologies fully integrated according to the transformation of the AGV products and RFID are derived from aerospace military technology. At present, the AGV products have been extended to aerospace, aviation, rail transit, intelligent logistics and other fields, forming a number of demonstration applications, solving the technical bottlenecks of precision machining, automated transfer, assembly and other large structures.

References

1. Wu, Q.: Present situation and developing trend of AGV key technology. Manufact. Autom. **35**(5), 106–107 (2013)
2. Wu, W.: Application status and development trend of AGV autonomous guided robot. Robot Tech. Appl. **25**(3), 16–17 (2012)

3. Hu, C.H.: Investigation of idle vehicle prepositioning strategies in an automated guided vehicle system, Ph.D. Dissertation: Department of Industrial and Manufacturing Engineering, The Pennsylvania State University, University Park, PA (1995)
4. Agullo, J., Cardona, S., Vivancos, J.: Dynamics of vehicles with directionally sliding wheels. Mech. Mach. Theor. **24**(1), 53–60 (2014)
5. Su, Y.: The omnidirectional mobile AGV research. Manufact. Autom. **36**(8), 10–11 (2014)
6. Xu, W., Cai, R.: AGV assembly robot based on IPC control system design. Appl. Electron. Tech. **39**(7), 131–134 (2013)
7. Xu, H., Zhao, Y.: Study on center line extraction for AGV's GuidRibbon under complex conditions. Comput. Measur. Control **39**(7), 131–134 (2013)
8. Meng, W., Liu, Z.: Research on visual guided AGV path tracking control. Control Eng. **21** (3), 321–325 (2014)
9. Liu, G., Guo, W.: Application of improved arithmetic median filtering denoising. Comput. Eng. Appl. **46**(10), 187–189 (2010)
10. Bonin-Font, F., Ortiz, A., Oliver, G.: Visual navigation for mobile robot: a survey. J. Intell. Rob. Syst. **53**(3), 263–296 (2008)
11. Yun, L., Xun, D.: Image segmentation based on gray equalization and improved genetic algorithm. Comput. Eng. Appl. **47**(16), 194–197 (2011)
12. Bhanu, B., Lee, S., Ming, J.: Adaptive image segmentation using a genetic algorithm. IEEE Trans. Syst. Man Cybenetics **25**(12), 1543–1546 (1995)
13. Ruan, Q., Ruan, Y.: Digital Image Processing, 2nd edn. Beijing Electronic Industry Publishing House, Beijing (2003)
14. Lee, J.W., Kim, J.H., Lee, Y.J., Lee, K.S.: A Study on recognition of road lane and movement of vehicles for port AGV vision system. In: Proceedings of the 40th SICE Annual Conference International Session Papers, SICE 2001 (2001)
15. Zhang, H.: Digital Image Processing Pattern Recognition Technology and Engineering Practice. People's Posts and Telecommunications Publishing House, Beijing (2003)
16. Cai, J.: Development and research of automatic vision guided vehicle. Postdoctoral research report, Zhejiang University (2007)

Improved Image Retrieval Algorithm of GoogLeNet Neural Network

Jun Zhang[(✉)], Yong Li, and Zixian Zeng

Engineering University of PAP, Weiyang District Xi'an 710086, China
2010396767@qq.com

Abstract. How to search images people are interested in quickly and accurately among large-scale image and video data becomes a challenge. The proposed deep learning combined with the hash function retrieval method can not only learn the advanced features of the image to eliminate the need to manually design the feature extractor, but also reduce dimension to improve the retrieval effect. Firstly, improve GoogLeNet neural network to extract the high-level features of pictures. Secondly, add the hash function hidden layer at different levels of the network to make up for the lack of detailed information of high-level features, and fuse the different features of the image to generate hash binary encoding with certain weights. Finally, the images are sorted by similarity according to the Hamming distance and the method would return similar images. The experimental results show that the proposed method has a significant improvement in accuracy and efficiency on the CIFAR-10 and NUS-WIDE datasets.

Keywords: Convolutional neural network · Hash learning · Feature fusion · Image retrieval

1 Related Work

The traditional Content-Based Image Retrieval (CBIR) algorithm uses computer to establish the image feature vector description and store it in the image feature database. When the user inputs a query image, computer would extract the feature representation and calculate the similarity in the feature library under a certain similarity measure. The return images are sorted according to the similarity size. And computer output sequentially the sort image [1]. Classic image retrieval algorithms achieve good results on small and middle datasets such as Bag of Words algorithm (BOW) [2], Vector of Locally Aggregated Descriptors (VLAD) algorithm [3], and Fisher Vector algorithm (FV) [4]. However, the access of multiple intelligent terminals and the development of mobile Internet make pictures and video become the most mainstream information. When the pictures are expanded to millions or even larger, the handcrafted filters consume a lot of manpower and resources. The semantic annotation is not clear enough, the output feature latitude is high, and the retrieval response has certain real-time requirements. The traditional algorithm can't meet the current wide-ranging requirements. People need an efficient and accurate algorithm to apply to large-scale image retrieval. Because of deep learning [5], especially the convolutional neural

© Springer Nature Switzerland AG 2020
J. H. Kim et al. (Eds.): ICHSA 2019, AISC 1063, pp. 25–34, 2020.
https://doi.org/10.1007/978-3-030-31967-0_3

network (CNN) is invariant to recognize the translation and rotation of the picture, the convolutional layer learns the picture and transforms it into a high-level feature representation familiar to human eyes, requiring only a small amount of manual intervention. It is very suitable for large-scale image retrieval and massive data.

Many researchers try to combine deep learning and hash function [6] in the field of image retrieval. Pan et al. [7] first propose encode image to binary vector as CNN input and learn a good feature representation for the input images as well as a set of hash functions. Lai et al. [8] regard the triplet image as input and use the divide-and-module module to slice the image with q branches. Lin et al. [9] propose to add a fully connected layer between fc6 and fc7 to generate a binary hash for image retrieval. In [10, 11], they input a pair of pictures for training and design carefully the loss function to output a binary code to retrieval image. Zhang et al. [12] propose multi-scale input and design a loss function, which is composed of cross entropy, L2 norm and balance term, to ensure the error and balance of hash coding. In this paper, we propose a method that combine GoogLeNet with hash function and learn and fuse the characteristics of different levels of images. Then join the hash function hidden layer after different levels of the network and encode the feature representation into a set of binary codes. Lastly sort the similarity according to the Hamming distance and return the most similar picture.

2   Improved GoogLeNet Neural Network

2.1   GoogLeNet Neural Network

GoogLeNet [13], a 22-layer-deep network, is the model structure adopted by the winner of the ILSVRC group on ImageNet in 2014. It not only improves the depth and width of the network, but also maintains a good representation of the image. And the model parameters are only 1/12 of the AlexNet. This paper is based on the original GoogLeNet to better identify and extract image features.

GoogLeNet is a classic convolutional neural network, consisting of input layer, convolutional layer, pooling layer, fully connected layer and output layer. The input layer directly inputs the preprocessed picture. The convolution layer connects the local area of the input picture to the neurons of the filter and extracts the features of the picture as input to the pooling layer by convolution and biasing. Its calculation formula is as follows:

$$x_{ij}^{xy} = \text{ReLu}\left(\sum_m \sum_{p=0}^{P_i-1} \sum_{q=0}^{Q_i-1} w_{ijm}^{pq} x_{(i-1)}^{(x+p)(y+q)} + b_{ij}\right) \tag{1}$$

ReLu($\bullet$) represents the Relu activation function. w_{ijm}^{pq} represents the connection weight of the convolution kernel and the m-th feature map of the previous layer. P_i and Q_i represents the width and height of the picture. Behind the convolutional layer is the pooling layer. It can make the feature map smaller and simplify the computational complexity of the network. Its calculation formula is as follows:

$$\partial_j^l = \beta_j^l \text{down}\left(\partial_j^{(l-1)}\right) + b_j^l \qquad (2)$$

Down(•) represents the down sampling function, generally adopting the max pooling or average pooling. They improve the robustness of the model by reducing the resolution of the previous feature map. In this paper, we use max pooling to select the maximum value of the image area as the pooled value of the area. β_j^l indicates the weight of the connection, b_j^l is the offset after pooling. The fully connected layer classifies the advanced feature representation. In order to prevent overfitting, CNN use the dropout function to randomly swap the weight of some of the connections to zero.

When the network goes deeper, the model parameters will increase sharply and it is more prone to over-fitting. The exploding parameters also require a large amount of computing resources. The most straightforward way to solve these two problems is to use sparse connection instead of full connection. By analyzing the correlation of certain activation values, the highly correlated neurons are aggregated to obtain an image representation. Inception structure is used to find the optimal local sparse structure and cover it as an approximate dense group. It adopts the parallel convolution method to jointly filter the output of the 1×1, 3×3, 5×5 convolution kernel and the 3×3 max pooling. As shown in Fig. 1.

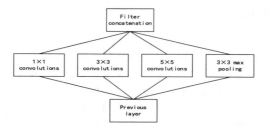

Fig. 1. Inception module

However, the 3×3 and 5×5 convolution kernels will cause the parameter quantity too large. Therefore, the 1×1 convolution kernel is added before the 3×3, 5×5 convolution kernel and behind 3×3 max pooling. As shown in Fig. 2, it not only reduces the parameter amount by dimension reduction but also increases the depth and width of the network.

GoogLeNet replace the full connection layer with the average pooling layer to reduce the model parameter and computational complexity. And it joins the auxiliary loss layer at the different level of the network and adds the loss to the total loss according to the discounted weight (discount weight of 0.3) (Fig. 3).

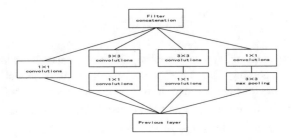

Fig. 2. Inception module after dimension reduction

2.2 Hash Function Layer

The probability that adjacent pictures can be mapped to close locations in the new dataspace is high, and the probability that different pictures are mapped to close positions is low:

If $d(x, y) \leq d_1$, the probability of $h(x) = h(y)$ is at least p_1.

If $d(x, y) \geq d_2$, the probability of $h(x) = h(y)$ is at most p_2.

$d(x, y)$ represents the distance between x and y and d_1 is less than d_2. $h(x)$ and $h(y)$ represent the hash transformation of x and y. GoogLeNet extracts the high-dimensional feature vector of the image through the average pooling layer. After learning by the hash function, it is mapped into a low-dimensional projection vector.

$$V_i = AverPooling\left(ReLu\left(W^T \emptyset(x_i; \theta) + bias_i\right)\right) \tag{3}$$

θ is the weight of each layer in the feature learning part. Φ is the eigenvector after sample learning. W is the weight matrix of the hash function layer and the average pooled layer. $bias_i$ is the offset vector. Then use the Sgn function to encode low-dimensional vectors and we can obtain a sets of binary code values.

$$b_i = sgn(V_i) \tag{4}$$

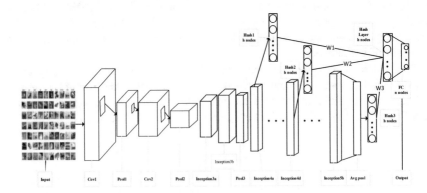

Fig. 3. Improve the overall architecture of GoogleNet neural network

In this paper, we add a hash function hidden layer in front of the GoogLeNet loss layer, as shown in Table 1, to perform hash function learning on the high-dimensional image features extracted by the neural network.

Table 1. GoogLeNet neural network and it's improvement

GoogLeNet	Improve GoogLeNet
Convolution	Convolution
Max pool	Max pool
Convolution	Convolution
Max pool	Max pool
Inception (3a–3b)	Inception (3a–3b)
Max pool	Max pool
Inception (4a–4e)	Inception (4a–4e)
Max pool	Max pool
Inception (5a–5b)	Inception (5a)
Avg pool	Avg pool
Softmax	Hash layer
	Softmax

The constructed hash function ensures the consistency of the high-dimensional image feature space and the projection vector space, and also ensures that each hash code carries the original image information as much as possible [14]. After the generated m-dimensional projection vector is quantized, the corresponding binary encoding is obtained. Discrete hash coding is subject to discrete conditions, and hash coding learning is difficult to optimize. Therefore, we adopt the Sigmoid function to relax the discrete conditions. It relaxes the hash code to the real number field, sets the threshold and binarizes the hash code.

$$b_i = \text{sigmoid}(V_i) \tag{5}$$

In order to further improve the retrieval performance, we extract the different levels of features and add corresponding hash function hidden layers in different loss layers. Then set the threshold and encode low-dimensional vectors to obtain binary code values. Lastly different weights are combined with binary code values obtained by encoding different levels of features, and add them up to obtain the final coding vector. When user input query image, computer would compute the Hamming distance, match the binary code library on the dataset and return the same or similar pictures.

3 Experimental Methods and Steps

3.1 Dataset

The CIFAR-10 dataset consists of 60,000 32 × 32 color images in 10 classes, with 6,000 images per class. There are 50,000 training images and 10,000 test images. The NUS-WIDE dataset contains 269,648 images collected from Flickr. The associations between images and 81 concepts are manually annotated. We use the images associated with the 21 most frequent concepts, where each of these concepts associates with at least 5,000 images, resulting in a total of 195,834 images. We also compare the traditional classic image hash algorithm with the deep learning and hash function. such as Supervised Hashing with Kernels (KSH) [15], Iterative Quantization (ITQ) [16] and other deep hash algorithm.

3.2 Experimental Methods

Step 1. Data enhancement and pre-processing. Convert the downloaded data to LMDB format. Because the image of some categories in the NUS-WIDE dataset is insufficient, we flip, pan, crop, and zoom the image.

Step 2. Download the GoogLeNet neural network model that has been pre-trained on ImageNet.

Step 3. Add corresponding hash function hidden layers in different loss layers and fine-tuning on the target data set for our experimental needs. We can obtain a model file containing image high-dimensional features and hash binary encoding.

Step 4. Extract the image high-dimensional features and the hash binary code in the model file. The query image is first barreled by hash binary code and return similar pictures with a certain Hamming distance. Then calculate the Euclidean distance between the high-dimensional features of the image in the same barrel and return Top K similar images.

Step 5. Analyze experimental results and compare with classic image hash algorithm and deep hash algorithm

4 Experimental Results and Performance Analysis

4.1 Performance Indicators

The image retrieval commonly uses accuracy and recall indicators to evaluate the search results. This is better in the single-label dataset, but in reality, multiple types of objects often coexist in one image. In order to better evaluate the performance of model retrieval, we use mAP (mean average precision) to compare the performance of different models in the target dataset. mAP is the average of multiple APs, and AP is the area enclosed by the P-R curve. In addition, we evaluate the ranking of top k images with respect to a query image q by a precision:

$$\text{Precision@k} = \frac{\sum_{i=1}^{k} Rel(i)}{k} \tag{6}$$

where $Rel(i)$ denotes the ground truth relevance between a query q and the i-th ranked image.

4.2   Results on CIFAR-10

Compared with the traditional hash algorithm, the performance of the deep hash algorithm has been greatly improved. As shown in Table 2, our method named GH add only one hash function hidden layer after average pooling layer. And it is about 2% higher than the current best method (DLBHC). It is also increased by 20% to 30% compared with other deep hash algorithm. On the basis of the previous, we add corresponding hash function hidden layers after two auxiliary loss layers and average pooling layer to generate binary code in certain weights (G3H). The G3H is about 1% higher than the DLBHC, but is about 1% lower than the GH. This is because in the preprocess of image datasets when we resize 32 * 32 to 256 * 256, the low-level feature information of the image has been lost because of image dilation. It leads that the hash code achieves poor results and affects the final retrieval performance.

Table 2. Comparison of mAP sorted by Hamming distance on the CIFAR-10 dataset

	12bit	24bit	32bit	48bit
G3H	0.8968	0.8983	0.8981	0.9008
GH	0.9029	0.9042	0.9063	0.9099
DLBHC [9]	0.875	0.882	0.891	0.894
DPSH [11]	0.713	0.727	0.744	0.757
DSH [10]	0.6157	0.6524	0.6607	0.6755
NINH [8]	0.552	0.566	0.558	0.581
CNNH [7]	0.489	0.511	0.509	0.522
KSH [15]	0.303	0.337	0.346	0.356
ITQ [16]	0.162	0.169	0.172	0.175

We can see the accuracy curve for each method of returning different numbers of similar pictures in Fig. 4. As the number of similar pictures returned increases, our method achieves the best performance. And the accuracy gradually increases and tends to be stable and both reach over 90%. Especially the accuracy of the GH algorithm reaches 91%. These curves show that the improved GoogLeNet algorithm not only improves the feature representation but also uses the image tags to make the binary encoding more preserve the similar information between the images, which significantly improves the accuracy of the retrieval.

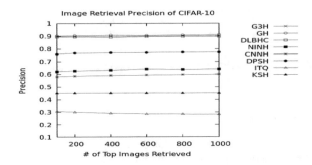

Fig. 4. The 48-bit search returns the top-1000 near-neighbor accuracy curve

4.3   Results on NUS-WIDE

As shown in Table 3, the method we proposed on the NUS-WIDE dataset also achieves relatively good performance. Compared with the best performing deep hash algorithm, the mAP value of our algorithm has increased respectively 0.55% and 1.71%. In particular, the G3H algorithm, which combines the high-level features and the middle-level features, performs outstanding and reaches 83%. This shows that our idea is feasible, and it combines the different levels of features to make up for the lack of detailed information on the high-level features. Compared to the GH algorithm that only extracts the high-level features, the mAP of the G3H algorithm is increased by about 2%, and the accuracy and efficiency are steadily improved.

Table 3.   Comparison of mAP of top-5000 neighbors on the NUS-WIDE dataset

	12bit	24bit	36bit	48bit
G3H	0.7712	0.7925	0.8186	0.8291
GH	0.7836	0.8092	0.8012	0.8175
DLBHC [9]	–	–	–	–
DPSH [11]	0.752	0.790	0.794	0.812
DSH [10]	0.5483	0.5513	0.5582	0.5621
NINH [8]	0.674	0.697	0.713	0.715
CNNH [7]	0.611	0.618	0.625	0.608
KSH [15]	0.556	0.572	0.581	0.588
ITQ [16]	0.452	0.468	0.472	0.477

Figure 5 shows the accuracy curve for returning 1000 images of hash code at different lengths. As the length of the hash code increases, the high-level features of the picture are more retained and the accuracy of the similarity matching query is improved. Compared with the accuracy of the 48-bit GH algorithm, the accuracy of the 12-bit G3H algorithm is very close. This also shows that the feature representation of the fusion with different hierarchical features indicates that hash learning is more representative and compact for binary coding. The accuracy of the G3H low-order encoding has been comparable to that of the GH high-order encoding. The storage space is greatly reduced and the retrieval time is rapidly reduced.

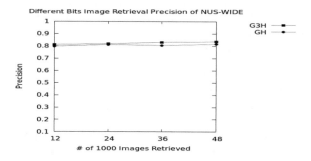

Fig. 5. Different hash coded digits retrieve returns 1000 image accuracy curves

5 Conclusion

In this paper, we propose improved GoogLeNet learning picture feature and the hash function generating binary code to reduce the storage overall and improve the retrieval efficiency. Improve the GoogLeNet neural network structure, add a hash function hidden layer after the average pooling layer and add hash function hidden layers at two auxiliary loss layers and average pooling layer. When learning advanced features, it can make up for the details of the picture and fuse the different levels of the picture to improve the picture feature expression. The experimental results show that the accuracy and efficiency are improved on the target dataset.

References

1. Li, T.: Deep hashing method based multi-table image retrieval. Beijing University of Posts and Telecommunications, Beijing (2018)
2. Aptoula, E.: Bag of morphological words for content-based geographical retrieval. In: International Workshop on Content-based Multimedia Indexing, Klagenfurt, pp. 1–5. IEEE Press (2014)
3. Jegou, H., Douze, M., Schmid, C.: Aggregating local descriptors into a compact image representation. In: IEEE Conference on Computer Vision and Pattern Recognition, Portland, Oregon, USA, pp. 3304–3311. IEEE Press, Piscataway (2010)
4. Sanchez, J., Perronnin, F., et al.: Image classification with the Fisher vector: theory and practice. Int. J. Comput. Vis. **105**(3), 222–245 (2013)
5. Krizhevsky, A., Sutskever, I., Hinton, G.E.: ImageNet classification with deep convolutional neural networks. In: Advances in Neural Information Processing Systems, pp. 1097–1105 (2012)
6. Tang, B., He, H.: Extended nearest neighbor method for pattern recognition. Comput. Intell. Mag. IEEE **10**(3), 52–60 (2015)

7. Xia, R.K., Pan, Y., Lai, H.J., et al.: Supervised hashing for image retrieval via image representation learning. In: Proceedings of the 28th AAAI Conference on Artificial Intelligence, Québec, pp. 2156–2162. AAAI Press (2014)

8. Lai, H.J., Pan, Y., Liu, Y., et al.: Simultaneous feature learning and hash coding with deep neural networks. In: IEEE Conference on Computer Vision and Pattern Recognition, Boston, pp. 3270–3278. IEEE Press (2015)

9. Lin, K., Yang, H., Hsiao, J.H., et al.: Deep learning of binary hash codes for fast image retrieval. In: IEEE Conference on Computer Vision and Pattern Recognition Workshops, Boston, pp. 27–35. IEEE Press (2015)

10. Liu, H., Wang, R., Shan, S., et al.: Deep supervised hashing for fast image retrieval. In: IEEE Conference on Computer Vision and Pattern Recognition, Las Vegas, pp. 2064–2072. IEEE Press (2016)

11. Li, W.J, Wang, S., Kang, W.C.: Feature learning based deep supervised hashing with pairwise labels. In: The 25th International Joint Conference on Artificial Intelligence, New York, pp. 1711–1717. AAAI Press (2016)

12. Zhang, Y.H., Huang, Z.C., Chen, Y.X.: MDBH: a multi-scale balanced deep hashing method for image retrieval. Chin. J. Appl. Res. Comput. **03**, 1–7 (2019)

13. Szegedy, C., Liu, W., Jia, Y., et al.: Going deeper with convolutions. In: IEEE Conference on Computer Vision and Pattern Recognition, Boston, pp. 1–9. IEEE Press (2015)

14. Norouzi, M., Fleet, D.J.: Minimal loss hashing for compact binary codes. In: Proceedings of the 28th International Conference on Machine Learning, Washington, pp. 353–360. International Machine Learning Society Press (2011)

15. Liu, W., Wang, J., Ji, R., et al.: Supervised hashing with kernels. In: IEEE Conference on Computer Vision and Pattern Recognition, Providence, pp. 2074–2081. IEEE Press (2012)

16. Gong, Y., Lazebnik, S.: Iterative quantization: a procrustean approach to learning binary codes. In: IEEE International Conference on Computer Vision and Pattern Recognition, Providence, pp. 817–824. IEEE Press (2011)

A Review of Face Anti-spoofing and Its Applications in China

Bizhu Wu, Meng Pan, and Yonghe Zhang[✉]

Shenzhen University, Shenzhen, China
yhzhang@szu.edu.cn

Abstract. Face anti-spoofing detection gains more and more attention, and due to some attacks like spoofing image, spoofing video and 3D masks etc. in security sensitive area, it has being widely used in industries in China. In this work, we summarize the traditional and recent methods proposed in the area of face anti-spoofing and divide them into three main categories, which are feature based methods, deep learning based methods and other methods. In addition, we also compare the performance of these methods and investigate the application of some famous Chinese enterprises. Finally, we provide an outlook into the future of this field of research.

Keywords: Biometrics · Liveness detection · Face anti-spoofing · Computer vision

1 Introduction

Face anti-spoofing is to determine whether the captured face is a real face or a forged face like a face color-printed in paper, a face image on the screen of digital device, 3D face masks, etc. Face anti-spoofing is often combined with face recognition system and face authentication system, aiming to verify whether the user is truly himself/herself, which have been widely used in the area of online payment, attendance registration system and other face authentication areas. Face anti-spoofing can be considered as a classification problem in computer vision area. In general, its algorithms can be divided into two types: binary classification problem, which aims to classify a face as genuine or spoofing, and multi classification problem, which task is to classify the type of spoofing attack as image spoofing, device screen spoofing, 3D mask spoofing, etc.

2 Current Research State of Face Anti-spoofing

2.1 Feature Engineering Based Approaches

In the early stage, traditional face anti-spoofing algorithms utilize hand-crafted features, finding out the difference between genuine face and spoofing face, and use algorithms to establish classifiers according to these features. The known differences between genuine face and spoofing face are color texture, non-rigid motion deformation,

© Springer Nature Switzerland AG 2020
J. H. Kim et al. (Eds.): ICHSA 2019, AISC 1063, pp. 35–43, 2020.
https://doi.org/10.1007/978-3-030-31967-0_4

materials (like skin, paper and mirror), quality of image and video. So, the literature of this stage is targeted to craft fixed features.

In Wen et al. [1], it uses a single frame as input and crafts statistical features such as mirror reflection, image quality distortion and color texture. After merging these features, it is directly trained a SVM Support Vector Machine (SVM) classifier to do the binary classification. However, it is hard to classify correctly when giving the model high-definition color-printed face images or high-definition face videos which have a low quality distortion.

In Boulkenafet et al. [2], a robust model is proposed by Oulu CMVS. It is based on an observation that the textures of genuine and spoofing images are more difficult to distinguish in RGB channels, but they are significantly different in other color spaces like HSV and YCbCr. The model utilizes multilevel LBP (Local Binary Pattern) Features of human face in HSV (Hue Saturation Value) space and LPQ (Local Phase Quantization) features (developed into SURF, Speeded-Up Robust Features, later) of human face in YCbCr space. It is proved that genuine and spoofing face are discriminative in other spaces such as HSV, and thus some following deep learning methods have used HSV channel or others as input to improve performance.

In Bharadwaj [3] and Tirunagari et al. [4], they used continuous multi frames of human face as input and crafted the features by finding the difference of micro-motion between genuine and spoofing faces. The Samarth Bharadwaj enhances micro-motion in the faces by motion magnification, and then extracts HOOF features and dynamic texture LBP-TOP features. Tirunagari et al. [4] obtains the subspace mapping of the maximum motion energy by DMD, and then analyzs the textures. However, the preprocessing of the motion magnification has a poor effect due to adding a lot of other band noise. This motion-based method is ineffective for spoofing face in wrapped paper jitters and video attack.

Li et al. [5]. proposed measuring heart rate in the videos containing faces in CVPR2014 and first applied remote pulse in the face anti-spoofing detection. They use pules distribution in different frequency domain and a LBP texture classifier to distinguish the genuine face and video attacks. It is a breakthrough to introduce the new mode of psychological signal. Meanwhile, it can also uses different deployment of classifier and corresponding features targeting at each type of spoofing attack. However, since the remote heart rate algorithm is not robust, the model based on pulse-feature is not reliable. Furthermore, it still needs to verify whether there exists difference in pulse features between the face in the video and the genuine one.

In Arora et al. [17], a face vitality detection algorithm with multiple activity indices is proposed. Blinking sequences, lip and chin movements are multiple living indicators that have been considered for reliable facial activity testing. Experiment on Eyeblink Dataset results show that the security of the face recognition system is improved significantly by this method combined with multiple activity indexes. The method achieves a higher activity detection rate by detecting photo attack, eye photo impersonation attack and video attack.

2.2 Deep Learning Based Approaches

Since 2015, people have already started using deep learning to study face anti-spoofing technology. However, due to limited training samples on open datasets, the performance of using deep learning can't go beyond the traditional method.

In Xu [6], it tries to use multi frames as input of the CNN-LSTM network to imitate the traditional method, but perform poorly. In Souza [7], it used single frame as input, segmented the face image, went through the pre-train network and then got a fine-tune in the whole face image. But it isn't very useful. Atoum et al. [8] is the first people considered the depth map as the difference between the genuine and the spoofing one. The spoofing face in the screen is generally flat, and even if the spoofed face in the paper is distorted, it is still different from the space distribution of the genuine faces. Even if they use a lot of tricks to fusion the model, performance still can't go beyond the traditional approach (see Fig. 1).

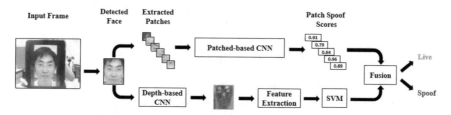

Fig. 1. Architecture of the proposed face anti-spoofing approach.

In Liu [9], they designed a deep learning model, using end-to-end learning approach to predict statistics of pulse and depth map. It didn't connect the final classifier layer in the model but directly made a threshold decision through the similar distance of the features of samples. Its performance finally outdo the traditional methods. It turns the binary classification problem into a target-oriented supervised learning feature problem, guaranteeing that the model was just learning these two features. They design Non-rigid Registration Layer to align the non-rigid motion of each frame (such as posture, expression, etc.), and then to learn the temporal pulse information better through RNN (see Fig. 2).

In Xiao et al. [10], Arxiv's biggest contribution is putting the face anti-spoofing detection directly into the face detection model (SSD, MTCNN, etc.) as a class. The result of this models were divided into background, genuine face and spoofed face, so it can filter out some spoofed attack at an early stage. So the speed of the whole system is very fast, which is very suitable for industrial deployment.

In Jourabloo et al. [11], the genuine face is seen as the original image, and the spoofed face image is regarded as the distortion one after the noise is added. So they estimates spoof noise, uses the Noise Pattern feature to classify, uses Encoder-Decoder to get the spoof noise N, and then reconstruct it by residuals. In order to ensure that the network is effective for different input and the learned noise is valid, they designed three loss functions, which are Magnitude Loss, Repetitive Loss and 0/1 Map Loss, to constrain the network according to the prior knowledge. It visualizes the Spoofing

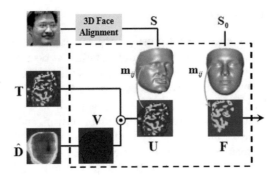

Fig. 2. The non-rigid registration layer.

Noise. But it is hard to deploy in actual scene because when the quality of spoofing attack is relatively high, Spoofing Noise can't perform well (see Fig. 3).

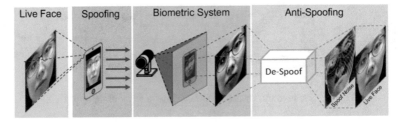

Fig. 3. The illustration of face spoofing and anti-spoofing processes. De-spoofing process aims to estimate a spoof noise from a spoof face and reconstruct the live face. The estimated spoof noise should be discriminative for face anti-spoofing.

In, Nagpal et al. [16] evaluates the performance of face anti-spoofing in CNN. The study uses Inception and ResNet CNN framework, and the results are calculated based on MSU Mobile Face Spoofing Database. The experiment is completed by considering the depth of the model, random weight initialization, weight transfer, fine-tuning and training from scratch, and different learning rates. Ultimately, these CNN models perform well in facial anti-spoofing in different environments.

2.3   Other Approaches

Near Infrared (NIR). Since the spectral band of the NIR is different from the visible light VIS, the absorption and reflection intensity of the genuine face and the spoofing face for the near-infrared band are different. So we can use the near-infrared camera to detect spoofing face. The near-infrared image on the video attack is very different from the genuine one, but the near-infrared image on the high-definition color paper printing is resemble to the genuine one. It can extract the light texture feature [12] or the remote face heart rate feature [13] in the NIR image. Using the same rPPG extraction method in Li et al. [5], the features shown in the NIR image are more robust.

Structure Light/ToF. Because structured light and ToF can reconstruct 3D face accurately in close distance, the point cloud and depth image of face and background can be obtained, which can be used as accurate face anti-spoofing detection. It is unlike the monocular RGB camera or binocular RGB camera which still need to estimate depth of image. However, the cost of this kind of reconstruction is high.

Light Field. The light field camera has an optical microscope lens array, and because the light field can describe the light intensity of any point in the space in any direction, the resulting raw light field photos and different refocusing photos can be used for in anti-spoofing detection. In Kim et al. [14], for the cheek edge of a real face in a microscope image, the pixels should be gradients with edges. However, for images from paper printing or screen attacks, the edge pixels are randomly and uniformly distributed. In Xie et al. [15], it is based on the theory that the depth information can be estimated from the difference between two refocusing images. As for feature extraction, the 3D face model of real face is different from that of spoofing face, and it can extract the brightness distribution feature, the sharpness of focusing image feature and spectrum histogram feature in different image.

2.4   Performance Comparisons

The performance of the face anti-spoofing detection algorithm are divided into two types: the internal testing of a single database and the cross-testing of multiple databases. As shown in the Table 1, the internal testing result of a single database [11]. As shown in the Table 2, cross-testing of multiple databases [11].

Table 1. The intra testing results on 4 protocols of Oulu-NPU.

Protocol	Method	APCER (%)	BPCER (%)	ACER (%)
1	CPqD [18]	2.9	10.8	6.9
	GRADIANT [18]	1.3	12.5	6.9
	Auxiliary [19]	1.6	1.6	1.6
	Face De-spoofing [11]	1.2	1.7	1.5
2	MixedFASNet [18]	9.7	2.5	6.1
	Face De-spoofing [11]	4.2	4.4	4.3
	Auxiliary [19]	2.7	2.7	2.7
	GRADIANT	3.1	1.9	2.5
3	MixedFASNet	5.3 ± 6.7	7.8 ± 5.5	6.5 ± 4.6
	GRADIANT	2.6 ± 3.9	5.0 ± 5.3	3.8 ± 2.4
	Face De-spoofing	4.0 ± 1.8	3.8 ± 1.2	3.6 ± 1.6
	Auxiliary [19]	2.7 ± 1.3	3.1 ± 1.7	2.9 ± 1.5
4	Massy-HNU [18]	35.8 ± 35.3	8.3 ± 4.1	22.1 ± 17.6
	GRADIANT	5.0 ± 4.5	15.0 ± 7.1	10.0 ± 5.0
	Auxiliary [19]	9.3 ± 5.6	10.4 ± 6.0	9.5 ± 6.0
	Face De-spoofing [11]	5.1 ± 6.3	6.1 ± 5.1	5.6 ± 5.7

Table 2. The HTER of different methods for the cross testing between the CASIA-MFSD and the Replay-Attack databases.

Method	Test	Train	Test	Train
	CASIA MFSD	Replay attack	CASIA MFSD	Replay attack
Motion [20]	50.2%		47.9%	
LBP-TOP [20]	49.7%		60.6%	
Motion-Mag [21]	50.1%		47.0%	
Spectral cubes [22]	34.4%		50.0%	
CNN [23]	48.5%		45.5%	
LBP [24]	47.0%		39.6%	
Colour texture [25]	30.3%		37.7%	
Auxiliary [19]	27.6%		28.4%	
Face De-spoofing [11]	28.5%		41.1%	

3   Applications of Face Anti-spoofing in China

3.1   Alibaba Cloud

Alibaba Cloud's interactive live detection uses the face feature point location tracking identification to reconstruct a resemble 3D faces. And it detects shaking, blinking, opening mouth and other actions to determine whether it is a genuine face. It carries out face detection, locates the feature and shapes 3D face model based on the CNN deep learning network to complete posture and motion estimation.

3.2   Tencent Cloud

Tencent Cloud's multidimensional live detection uses more complex random multi-digital lip language to capture the subtle changes in the mouth of people in the process of speech, so it is hard to forged a video. It also carries out voice-image synchronization detection, face texture analysis, mask detection, video anti-flip and other multidimensional protection means. Finally, all these tools are cross-fused, and realize the strong protection system which consists of mobiles and back-end.

3.3   Baidu Cloud

Baidu Cloud has six live detection technologies: motion-matched live detection, online picture live detection, H5 video live detection, offline RGB live detection, offline near-infrared live detection, offline 3D structured light live detection. For example, with the motion-matched live detection, users need to complete the specified action. System detects through real-time detection of the user's eyes, mouth, head posture status, to determine whether it is living. Offline 3D structured light live detection approach constructs depth maps by reflecting light from the surface of the genuine face to determine whether the targeted face is a genuine one, and it can effectively defend against pictures, videos, screens, molds and other spoofing attacks.

3.4 CloudWalk Technology

CloudWalk Technology's live detection is divided into software and hardware live detection, software live action detection can guide the user to complete the specified action within the specified time, to detect whether it is a genuine face. It has the advantages of high accuracy, convenience and fastness, while it can provide any business system an interface. The hardware live detection product is mainly an infrared binocular camera. It can simultaneously collect both near-infrared and visible pictures in real time to detect whether the target is a genuine face.

3.5 SenseTime

The face anti-spoofing detection of SenseTime takes a whole new approach, which the biggest difference from most of the traditional methods is that it pays more attention to the media properties of the spoofing attack, as well as the difference between the spoofed videos or images and the genuine videos or images.

4 Trends of Face Anti-spoofing

4.1 Method of 3D Reconstruction of Face Information

At present, the algorithm based on 3D face information is one of the best algorithms. Theoretically, all attacks based on flat media (such as printing paper, video) can be rejected by this methodology. However, methods based on this method are not fully developed, mainly because 3D reconstruction technology or 3D camera is still developing and popularizing. It can be expected that there will be more face anti-spoofing detection methods based on 3D human in future.

4.2 Methods Based on Multimodal Fusion by Various Sensors

Methods based on multimodal fusion gather and fuse the information from various sensors or various ways of the use of sensors. The information of a single sensor is insufficient in some cases, leaving attackers with opportunities to design unknown attacks. In order to avoid the limitation of different sensors, we can use the multi-sensors to carry out the face anti-spoofing detection in multi-dimensions. For example, the binocular camera with visible and infrared channels makes the algorithm use more spectral information to enhance the input data. Different sensors have different imaging features of facial skin, which is beneficial to detect 3D mask attack. In future, the new multi-sensors method will improve the performance of face anti-spoofing detection.

4.3 Methods Based on Generative Model

Most of current models are discriminative, letting algorithm learns the categorized labels directly from the feature space. However, the generative model instead learning the features of each category. Regarding the superior performance of the new Noise Model [11], the method based on generative models has very good potential.

Acknowledgement. This work is supported by two funding: the 13th Five-Year Plan of Education Science in Guangdong Province in 2017 (Contextual Moral Education Game Oriented to Collaborative Construction and Its Application, No. 2017JKDY43) and the Guangdong Postgraduate Education Innovation Project in 2018 (New Media and New Technologies in eLearning).

References

1. Wen, D., Han, H., Jain, A.K.: Face spoof detection with image distortion analysis. IEEE Trans. Inf. Forensics Secur. **10**(4), 746–761 (2015)
2. Boulkenafet, Z., Komulainen, J., Hadid, A.: Face spoofing detection using colour texture analysis. IEEE Trans. Inf. Forensics Secur. **11**(8), 1 (2016)
3. Bharadwaj, S., Dhamecha, T.I., et al.: Face Anti-spoofing via motion magnification and multifeature videolet aggregation (2014)
4. Tirunagari, S., Poh, N., Windridge, D., et al.: Detection of face spoofing using visual dynamics. IEEE Trans. Inf. Forensics Secur. **10**(4), 762–777 (2015)
5. Li, X., Komulainen, J., Zhao, G., et al.: Generalized face anti-spoofing by detecting pulse from face videos. In: 2016 23rd International Conference on Pattern Recognition (ICPR). IEEE (2016)
6. Xu, Z., Li, S., Deng, W.: Learning temporal features using LSTM-CNN architecture for face anti-spoofing. In: 2015 3rd IAPR Asian Conference on Pattern Recognition (ACPR). IEEE (2015)
7. Souza, G.B.D., Papa, J.P., Marana, A.N.: On the learning of deep local features for robust face spoofing detection (2018)
8. Atoum, Y., Liu, Y., Jourabloo, A., et al.: Face anti-spoofing using patch and depth-based CNNs. In: IEEE International Joint Conference on Biometrics. IEEE (2018)
9. Liu, Y., Jourabloo, A., Liu, X.: Learning deep models for face anti-spoofing: binary or auxiliary supervision (2018)
10. Xiao, S., Xu, Z., Liangji, F., et al.: Discriminative representation combinations for accurate face spoofing detection. Pattern Recognition S0031320318303182 (2018)
11. Jourabloo, A., Liu, Y., Liu, X.: Face de-spoofing: anti-spoofing via noise modeling (2018)
12. Sun, X., Huang, L., Liu, C.: Context based face spoofing detection using active near-infrared images. In: 2016 23rd International Conference on Pattern Recognition (ICPR). IEEE (2016)
13. Hernandez-Ortega, J., Fierrez, J., Morales, A., et al.: Time analysis of pulse-based face anti-spoofing in visible and NIR. In: 2018 IEEE/CVF Conference on Computer Vision and Pattern Recognition Workshops (CVPRW). IEEE Computer Society (2018)
14. Kim, S., Ban, Y., Lee, S.: Face liveness detection using a light field camera. Sensors **14**(12), 22471–22499 (2014)
15. Xie, X., Gao, Y., Zheng, W.S., et al.: One-snapshot face anti-spoofing using a light field camera (2017)
16. Nagpal, C., Dubey, S.R.: A performance evaluation of convolutional neural networks for face anti spoofing (2018)
17. Arora, A.S., Singh, M.: A novel face liveness detection algorithm with multiple liveness indicators. Wirel. Personal Commun. **100**(4), 1677–1687 (2018)
18. Boulkenafet, Z.: A competition on generalized software-based face presentation attack detection in mobile scenarios. In: ICJB. IEEE (2017)
19. Liu, Y., Jourabloo, A., Liu, X.: Learning deep models for face anti-spoofing: binary or auxiliary supervision. In: CVPR. IEEE (2018)

20. de Freitas Pereira, T., Anjos, A., De Martino, J.M., Marcel, S.: Can face anti-spoofing countermeasures work in a real world scenario? In: ICB. IEEE (2013)
21. Bharadwaj, S., Dhamecha, T.I., Vatsa, M., Singh, R.: Computationally efficient face spoofing detection with motion magnification. In: CVPRW. IEEE (2013)
22. Pinto, A., Pedrini, H., Schwartz, W.R., Rocha, A.: Face spoofing detection through visual codebooks of spectral temporal cubes. IEEE Trans. Image Process. **24**(12), 4726–4740 (2015)
23. Yang, J., Lei, Z., Li, S.Z.: Learn convolutional neural network for face anti-spoofing. Comput. Sci. **9218**, 373–384 (2014)
24. Boulkenafet, Z., Komulainen, J., Hadid, A.: Face anti-spoofing based on color texture analysis. In: ICIP. IEEE (2015)
25. Boulkenafet, Z., Komulainen, J., Hadid, A.: Face spoofing detection using colour texture analysis. IEEE Trans. Inf. Forensics Secur. **11**(8), 1818–1830 (2016)

An Event Based Detection of Internal Threat to Information System

Zheng Li[1] and Kun Liu[2,3]($\boxtimes$)

[1] School of Managerment, Hefei University of Technology, Hefei 230009, China
[2] University of Chinese Academy of Sciences, Beijing 100049, China
liukun315@mails.ucas.ac.cn
[3] Institute of Software, Chinese Academy of Sciences, Beijing 100190, China

Abstract. Internal threat is an important issue for the information systems of an organization. To deal with this problem, organizations often formulate regulations and rules to regulate the behavior of employees and prevent them from causing production risks. However, how to effectively detect violations of the rules in the production process is challenging. In this paper, we propose an event based internal threat detection method. Firstly, we establish a detection model for regulation violation by representing rules and regulations as complex events and design a rule engine to detect if these complex events occur and discover the violations of rules. Then the logs generated during product are used for activating the rule reasoning. Finally, the rule violation will be reported to the supervisor for further investigation. The experiment on the real production processes shows the method is effective and efficient to detect internal threats and can be used at major production sites.

Keywords: Internal threat · Event detection · Rule engine · Complex event

1 Introduction

With the widespread application of information system [1], organizations pay more and more attention to its security [2] and use various means to strengthen the security of information system. For example, using access control to limit the access to guard key places [3], building remote data center prevents single point of failure, etc. These measures effectively reduce external threats that may cause failures or business interruption [4]. On the other hand, misconduct, vandalism acts of internal staff with legitimate authority have become the greatest threat to the information system and one of the few security risks that may paralyze the entire business [5]. In addition, external attackers can act as candidates or services outsourcing to obtain legitimate internal identity for attacking or stealing secrets [6]. Therefore, the risk management of internal staff is an important part of the information system safety management.

© Springer Nature Switzerland AG 2020
J. H. Kim et al. (Eds.): ICHSA 2019, AISC 1063, pp. 44–53, 2020.
https://doi.org/10.1007/978-3-030-31967-0_5

In view of the possible safety problems caused by internal staffs, a series of rules and regulations are often formulated in the organization to prevent risks and control the process of possible accidents [7]. However, there are problems with violation detection of the rules and regulations [8]. For example, when the rules and regulations themselves are mostly descriptive texts, they easily lead to different detection standards; when the detection is performed manually, the regulation can result in inefficiencies and even become a mere formality. These problems lead to internal violations of rules and regulations that are difficult to detect and alert in time.

In such a situation, the behavior records of the employees operation on information system can be an efficient method to find out the internal threats [9]. These data can be used to characterize the behavior of internal staffs [10]. For example, the access control log shows the peoples position information; process approval data shows the operation of the process; the operation log records the actual operation and so on. In this paper, a complex event detection model is used to establish an internal behavior rule base based on rules and regulations and actual production, to detect various behavior data generated by internal staffs, to record and alarm violations of rules, to provide clues for the supervision department, identify potential safety hazards and reduce the number of production accidents.

2   Related Work

Internal security risks to information systems are an important issue. A series of global surveys of more than 1200 enterprises though 2013 to 2017 showed that internal attacks accounted for 25% or more of all threats (less than malware and phishing) and respondents thought the internal staff is the major vulnerability of information security (60%) [11].

To this end, researchers proposed a variety of different solutions. Hu [12] mapped Role-Based Access Control (RBAC) to security policies, used Genetic Algorithm (GA) to identify discrepancies between roles and process. Myers [13] used web server logs to compare and identify normal duty activity and suspicious internal activity. Spitzner [14] proposed to use honeypots to deal with internal attackers, especially to deal with new types of attacks, detect and track down attackers. Eberle [15] described graph-based approaches which they used to find out malicious internal activity in network-shaped domain such as social network, business processes etc. Rashid [16] used Hidden Markov Models to learn internal staffs normal behavior and then to detect deviations from normal. Stavrou [17] introduced an approach by adding external information evaluated from social media etc. to business process monitoring tools to pinpoint potential threaten insiders.

An event is defined as an occurrence of significance in as system [18], such as a connection to a system through a client, an approval of an operation, or swiping an ID card to enter a restricted area. A primitive event is the smallest, atomic event that happens completely or not at all [19]. A complex event (or a composite event) is calculated algebraically through simple or recursively, complex events [19]. We use primitive events to express the action of internal staffs and use complex events to represent rules and regulations.

3   Internal Threat Detection

3.1   Rules and Regulations

Rules and regulations are often descriptive text which is easy to read and understand by human, but not easy to understand by computer, for which we first need to formalize the rules and regulations of a unified representation. The rules and regulations essentially define the relationship between a specific behavior and the conditions it needs to meet. For example, in general, the changing operation of a production environment should be performed at a terminal in controlled area (of course, emergency will be exception). We use the representation of complex events, for each of the rules and regulations, we put it as a complex event, a rule describes a specific behavior, conditions to meet, relationship between the two expressed as a triple

$$(premise, conclusion, constraint)$$

Among them, premise (primitive or composite) event refers to the specific behavior which the rule and regulation concern (likely to lead to a security risk, such as production environment change operation in previous case); conclusion event (primitive or composite) are required conditions must be satisfied when premise occurs specified in rules and regulations (typically measures to guard against security risks, such as those that are required to be at a terminal in controlled area in the previous case); Constraints are the relationships that must be satisfied between premise and conclusion (for example, the time and position relationship between production environment change operation and the terminal in controlled area in the previous example).

3.2   Internal Staffs' Behavior

The process control formed in accordance with the rules and regulations truly recorded the internal staffs' behavior. As in the previous example, the access control system documented the process of staff entry into the controlled area and the operating system log recorded the staffs system login and operation. For the basic behavior of internal staff, refer to the rules and regulations expressed as a complex event, which can be expressed as a primitive event, such as access control system certification, operational application approval, system login, system operation. Each type of behavior is expressed as a primitive type of event. We have broken down internal staffs behavior records collected everywhere into primitive events and collectively expressed as

$$(staff, time, event, eventattribute)$$

The event attribute is an attribute unique to this type of event. For example, the attributes of the authentication event of the access control system include whether the authentication is successful, the location of the authentication device, the area controlled by the authentication device, or the like.

3.3   Detection Model

Combined with the rules and regulations as a complex event and represent internal staffs behavior as a primitive event, according to rules and regulations and internal staffs behavior common sense (unwritten rules, but prone to risk to fill regulatory loopholes), we choose the primitive events, event attributes and symbolic that meets regulatory compliance rules. In accordance with the rules and regulations of the formal representation, expressed as

$$(premise, conclusion, constraint)$$

The premise and conclusion events are made up of primitive events, logical operators (such as AND, OR, NOT) and the constraints consist of primitive event attributes and arithmetic operators (e.g. $>, <, =, etc.$). As in the previous example: "the changing operation of a production environment should be performed at a terminal in controlled area." The scenario described in this regulation is:

1. The staff enters into controlled area through access control system. *(primitive event: entering into controlled area)*
2. The staff logs into the production environment via the terminal in the controlled area. *(primitive event: login into production environment)* moreover, is not allowed to leave the controlled area from now on. *(primitive event: leaving controlled area or any action can be confirmed that the staff leaves controlled area such as re-enter from a non-controlled area into controlled area)*
3. The staff perform a series of operations *(each operation is a primitive event)*
4. The staff logs out of the production environment after completing the production operation *(primitive event: Log out of production environment)*, and can then leave the controlled area *(primitive event: exit from controlled area)*

For the above scenario, we extracted the basic events and main attributes: (Omit the personnel information and time information included in all the basic events, omit the production operations unrelated to the rules and regulations) (Table 1).

Table 1. Primitive event and attributes related to regulation

Primitive event	Short	Attributes
Access control system	AC	Pos_before
		Pos_after
Login into system	Login	System
Logout from system	Logout	System

The rules and regulations:

- Premise: $Login \land Logout$
- Conclusion: $AC\_1 \land \neg AC\_2$
- Constraints: (omitted for all events generated by the same person constraints)

   $((AC\_1.Pos\_after \in Controlled) \land (AC\_1.Time < Login.Time))$
   $\land(((AC\_2.Pos\_before \in Uncontrolled) \lor (AC\_2.Pos\_after \in Uncontrolled))$
   $\land(Login.Time < AC_2.time < Logout.time))$
   $\land((Login.System \in Production) \land (Logout.System = Login.System))$

Note that the rules in this example do not deal with situations where only the system is logged in and not logged out (intentional, negligent, or due to missing data). Such situation should be handled by other rules. (For example Premise: Login, Conclusion: Logout, Constraints: $Logout.Time - Login.Time \leq MaxOperTime$).

3.4   Detection

For a rule $(premise, conclusion, constraint)$, if the premise event does not occur, the rule is not triggered (not a violation), and the conclusion event is no longer detected; if the premise event occurs, the compliance rule is triggered to detect whether the corresponding conclusion event occurs. If it happens then it is in compliance with rules (compliance). If it does not happen, then it violates the rules (violation) (Table 2).

Table 2. Rule detection matrix

Premise	Conclusion	Result	Detection
Not occurs	Not occurs	$\neg Premise \land \neg Conclusion$	not a violation (not triggered)
Not occurs	Occurs	$\neg Premise \land Conclusion$	not a violation (not triggered)
Occurs	Not occurs	$Premise \land \neg Conclusion$	Violation
Occurs	Occurs	$Premise \land Conclusion$	Compliance

It is not a violation that the compliance rule is not triggered. Therefore, $Premise \land \neg Conclusion$ is a necessary and sufficient condition for violations. For violation detection, we do not care about the situation where the rules do not trigger ($\neg Premise \land \neg Conclusion$ and $\neg Premise \land Conclusion$) and behavioral compliance ($Premise \land Conclusion$). So, we only need to detect the behavior violation ($Premise \land \neg Conclusion$).

If the premise event and the conclusion event are composite events, use logical arithmetic operations to perform equivalent transformations to represent them as disjunctive normal form(DNF). For the above example (omit constraints)

- Premise: $Login \land Logout$
- Conclusion: $AC\_1 \land \neg AC\_2$

- We transfer premise event and conclusion event into DNF.

$$Premise \land \neg Conclusion \rightarrow Login \land Logout \land \neg(AC\_1 \land \neg AC\_2)$$
$$\rightarrow (Login \land Logout \land \neg AC\_1) \lor (Login \land Logout \land AC\_2)$$

We can notice that each conjunction expression is a violation of the rule, so we can use it to verify that the compliance rule covers all violations. Finally, we add up constrains to it, use rule engine to test if there are any compliance violations.

If the compliance rules themselves are too complicated, we can also use logical arithmetic operations to appropriately decompose a complex compliance rule into several sub-rules according to its DNF.

$$DNF : \sum \left(\prod_i p_i \cdot \prod_j \overline{c}_j \right) \Leftrightarrow \sum \left(\prod_i p_i \cdot \overline{\left(\sum_j c_j \right)} \right)$$

$$SubRule : \left(\prod_i p_i, \sum_j c_j, \mathcal{C} \right)$$

with p for Premise, c for Conclusion, $\mathcal{C}$ for Constraints

We use this to reduce the complexity of single rules and facilitate the understanding and maintenance of rules.

By this method, the rules and regulations are established as a compliant rule base, which serves as the basis for the detection of violations. When a rule needs to be detected, the rule is expressed as a violation determination formula, and convert into a disjunctive normal form. The internal staff who need to comply with the rule are judged one by one to confirm whether the violation occurs. If the violation occurs, the details of the violation are output and alarmed.

4   Evaluation

4.1   Production Evaluation

We deployed this platform at a bank's data center and set up a series of detection models (about 20) for the rules and regulations that are most relevant to system login. To test the effectiveness of the platform, we selected access control records (41), application and approve records (6) and login logs (12) from one of the internal staff (ID: AD13972) for one week (2016/11/28 to 2016/12/02). Conducted a non-compliance test on its production and maintenance.

The result is as follows: For all login behavior (12 times), violation occurred twice. The rules which were violated are shown in Table 3.

The first violation occurred in 2016/11/29 9:35:00 showed the staff did not get any approvement before login happened. Further investigation showed 9:30 am the same day, the production environment failed, the staff as major operation

Table 3. Violated rules in evaluation

Description	Premise	Conclusion	Constraints
Operation must be approved in advance	$Login$	$Approve$	$(Login.Time > Approve.Time)$ $\wedge(Approve.Estimated\_start < Login.Time$ $< Approve.Estimated\_end)$ $\wedge(Login.System = Approve.System)$ $\wedge(Login.OS\_USER = Approve.OS\_USER)$
Operation must be in controlled area	$Login$	AC	$(Login.Time > AC.Time)$ $\wedge(Login.Time - AC.Time \leq 1hour)$ $\wedge(AC.Pos\_after \in "Controlled")$

and maintenance personnel involved in emergency recovery, the resumption of operations by senior staff on-site supervision, operation approval made later following emergency protocol. This violation was not considered an internal threat.

The second violation occurred in 2016/12/2 19:02:00 shows the staff did not get any approvement before login happened and not in the controlled area while operating. Further investigation showed after the employee got off work and left the company, another employee in the same group used his ID to carry out production operations without any application due to the need of a temporary operation and did not report to senior. This violation was considered an internal threat due to improperly kept of ID and unauthorized use of others account.

After careful manual inspection, the employee only violated the rules this week twice in the same way as the test results. The above rule detection method can ensure 100% successful detection of the violation of the corresponding rules and regulations when the rules are set reasonably.

4.2   Production Deployment

After completing the production Evaluation, we applied the platform to the production environment. We periodically collect various data generated by internal personnel from various departments, including the access records of the access control system mentioned above, the login/logout log, the operation log of the production environment, and various application approval records, which are strictly related to production operations. We also obtained video surveillance records from some controlled areas from the security department (due to the lighting conditions and the camera angle, facial recognition is impractical, only the number of people can be identified), and other record related to the behavior of the production staff. Combining the recorded with the rules and regulations, we have established relevant detection models to detect whether there are violations. Such as combining the data of the access control record and the number of people in the control area in the monitoring record to determine whether anyone trailed with others entered the control area, which is a common violation with the access control system [20].

In addition to the rules and regulations, we will also sort out some behavioral regularity models concerning the operation and maintenance personnel's production operations to detect some situations that are not in violation of the rules and regulations but are worthy of attention. For example, in the model of the relationship between specific commands, we define the relationship between a series of commands. For example, after stopping the service, there should be an operation to start it; the execution of the command in a specific environment should enter into that environment first and so on. In the process of performing detection on such models, we noticed that some staff did not enter the SQL environment when executing SQL statements or SQL scripts. Some sample in Table 4.

Table 4. Missing SQL command

Date	Person	Operation	Note
2016/12/07 20:25:18	ZY	df -h su -oracle select * from dba_data_files exit	Missing sqlplus before select
2016/12/16 21:02:21	ZXJ	su -grid -c asmcmd lsdg su -oracle @show_tablespaces exit	Missing sqlplus before @show_tablespaces

Although these were merely misconduct caused by carelessness, and there is no production accident in the result, it is not a good sign in the production environment. We reflect these conditions to the production department and recommend that they should improve the processes and management mechanisms for the operation of the production environment.

5   Conclusion

In view of the implementation and detection of rules and regulations, this paper proposes a method of using complex event detection, formally formulating rules and regulations into complex events and effectively examining whether internal staffs behaviors conform to rules and regulations. Experiments show that under the condition of reasonable and complete rules and regulations, the internal abnormal behaviors and internal threats can be effectively detected, and the data generated by various internal controls and management are also effectively used.

However, this method is very dependent on the completeness of the rules and regulations. In practice, the rules and regulations often have loopholes. We also took this into consideration when we implemented the deployment, we added employees' routine behavior as common sense to improve the ability to detect internal threats.

On the other hand, this approach is essentially prescriptive compliance and therefore cannot be used to detect unknown threats. In future work, we consider using data mining, machine learning and other methods to analyze the employee's behavioral characteristics and infer the unknown internal threats that the rules and regulations did not take into account.

References

1. Bulgurcu, B., Cavusoglu, H., Benbasat, I.: Information security policy compliance: an empirical study of rationality-based beliefs and information security awareness. MIS Q. **34**(3), 523–548 (2010)
2. Dubois, E., Heymans, P., Mayer, N., Matulevičius, R.: A systematic approach to define the domain of information system security risk management. In: Intentional Perspectives on Information Systems Engineering, pp. 289–306. Springer, Berlin (2010)
3. Sandhu, R.S., Samarati, P.: Access control: principle and practice. IEEE Commun. Mag. **32**(9), 40–48 (1994)
4. Rivest, R.L.: U.S. Patent No. 4,376,299. U.S. Patent and Trademark Office, Washington, DC (1983)
5. Warkentin, M., Willison, R.: Behavioral and policy issues in information systems security: the insider threat. Eur. J. Inf. Syst. **18**(2), 101–105 (2009)
6. Colwill, C.: Human factors in information security: the insider threat who can you trust these days? Inf. Secur. Tech. Rep. **14**(4), 186–196 (2009)
7. Doherty, N.F., Fulford, H.: Aligning the information security policy with the strategic information systems plan. Comput. Secur. **25**(1), 55–63 (2006)
8. Herath, T., Rao, H.R.: Protection motivation and deterrence: a framework for security policy compliance in organisations. Eur. J. Inf. Syst. **18**(2), 106–125 (2009)
9. Bishop, M., Gates, C.: Defining the insider threat. In: Workshop on Cyber Security & Information Intelligence Research: Developing Strategies to Meet the Cyber Security & Information Intelligence Challenges Ahead, p. 15. ACM (2008)
10. Kandias, M., Mylonas, A., Virvilis, N., Theoharidou, M., Gritzalis, D.: An insider threat prediction model. In: International Conference on Trust, Privacy and Security in Digital Business, pp. 26–37. Springer, Berlin (2010)
11. van Kessel, P.: Cybersecurity regained: preparing to face cyber attacks, Ernst & Young Global Limited. Web. 17 November 2017. http://www.ey.com/Publication/vwLUAssets/ey-cybersecurity-regained-preparing-to-face-cyber-attacks/$FILE/ey-cybersecurity-regained-preparing-to-face-cyber-attacks.pdf
12. Hu, N., Bradford, P.G., Liu, J.: Applying role based access control and genetic algorithms to insider threat detection. In: ACM Southeast Regional Conference: Proceedings of the 44th Annual Southeast Regional Conference, vol. 2006, pp. 790–791, March 2006
13. Myers, J., Grimaila, M.R., Mills, R.F.: Towards insider threat detection using web server logs. In: Proceedings of the 5th Annual Workshop on Cyber Security and Information Intelligence Research: Cyber Security and Information Intelligence Challenges and Strategies, p. 54. ACM, April 2009

14. Spitzner, L.: Honeypots: catching the insider threat. In: Proceedings of the 19th Annual Computer Security Applications Conference, pp. 170–179. IEEE, December 2003
15. Eberle, W., Graves, J., Holder, L.: Insider threat detection using a graph-based approach. J. Appl. Secur. Res. **6**(1), 32–81 (2010)
16. Rashid, T., Agrafiotis, I., Nurse, J.R.: A new take on detecting insider threats: exploring the use of Hidden Markov Models. In: Proceedings of the 8th ACM CCS International Workshop on Managing Insider Security Threats, pp. 47–56. ACM, October 2016
17. Stavrou, V., Kandias, M., Karoulas, G., Gritzalis, D.: Business process modeling for insider threat monitoring and handling. In: International Conference on Trust, Privacy and Security in Digital Business, pp. 119–131. Springer, Cham, September 2014
18. Luckham, D.: The Power of Events, vol. 4. Addison-Wesley, Reading (2002)
19. Rizvi, S.: Complex event processing beyond active databases: streams and uncertainties. TR EECS-200526 (2005)
20. Norman, T.L.: Electronic Access Control. Elsevier, Burlington (2011)

Loop Closure Detection for Visual SLAM Using Simplified Convolution Neural Network

Bingbing Xu[1,2(✉)], Jinfu Yang[1,2], Mingai Li[1,2], Suishuo Wu[1,2], and Yi Shan[1,2]

[1] Faculty of Information Technology, Beijing University of Technology, Beijing 100124, China
S201602088@emails.bjut.edu.cn
[2] Beijing Key Laboratory of Computational Intelligence and Intelligent System, Beijing 100124, China

Abstract. Loop closure detection plays a vital role in visual simultaneous localization and mapping (SLAM), since it can reduce the accumulated errors. Handcrafted feature-based methods for loop closure detection have the weakness of lack of robustness with respect to illumination and scale changes. In recent years, the Convolutional Neural Networks (CNN) has been widely used in image recognition due to its feature expressive capacity. However, these deep networks are too complex to satisfy real-time for loop closure detection. In this paper, we propose a novel simplified convolutional neural network (SCNN) for loop closure detection in visual SLAM. We first perform superpixel processing on the original image to reduce the effects of illumination changes. In order to reduce the size of the parameters, we invert the feature maps obtained by convoluting and concatenate them with the original. In addition, we pre-train the proposed network on Places dataset to solve the problem of the scene being unlabeled. Experimental results over CityCenter and NewCollege dataset show that our proposed method can achieve better performance counterparts than other.

Keywords: Visual simultaneous localization and mapping (VSLAM) ·
Loop closure detection · Simplified convolutional neural network (SCNN)

1 Introduction

In recent years, the simultaneous localization and mapping (SLAM) is widely applied on robots, unman driving, augmented reality (AR) and virtual reality (VR), etc. SLAM is a highly active research topic in the field of computer vision and robot. The SLAM system can build a map of an unknown environment and localize the robot in the map on real-time operation [1]. A typical SLAM system consists of front-end feature extraction, back-end optimizing pose and mapping, and loop closure detection, among which the loop closure detection plays an important role in improving the accuracy and robustness of a whole SLAM systems.

The core problem of loop closure detection is scene recognition. Once the loops are detected correctly, the robot will relocate itself, adjust and optimize the global consist

© Springer Nature Switzerland AG 2020
J. H. Kim et al. (Eds.): ICHSA 2019, AISC 1063, pp. 54–62, 2020.
https://doi.org/10.1007/978-3-030-31967-0_6

map. For example, in the visual SLAM system based on graph optimization, the loops can reduce the cumulative error of the established environment map and improve the precision of back-end optimization.

The loop closure detection for visual SLAM is commonly used to find an identical scene, and in this process may occur the movement of scene objects and changes in lighting. The traditional visual loop closure detection algorithm registers the current position image with the previous position image. At present, the easiest way to detect loops is to match the visual features of all key-frames, for example, the bag-of-words (BOW) model [2], and the features are mostly handcrafted, such as: SIFT [3], SURF [4], ORB [5], and so on. But, this method has the weakness of lack of robustness with respect to illumination and scale changes. Because it ignores useful information about the environment.

Recently, deep learning and convolutional neural network (CNN) [6] have achieved good results in image recognition. This provides a new idea to solving loop closure detection problems. Generally, the framework of CNN composes of multi-layers and learns abstract features from input images, which is different from the traditional handcrafted features. In many of the current researches, the CNN's capacity to extract abstraction levels from visual data features has surpassed the performance of solutions based on handcrafted features. There are many studies that use CNN for loop closure detection, but these methods have to take a long time to train a CNN model. Therefore, it is very important to choose a simplified CNN model for loop closure detection and reduce the time of training and extracting features to meet the real-time requirements of visual SLAM.

In this paper, we propose a loop closure detection method based on the simplified convolutional neural network. We first performed superpixel processing on the original image to reduce the effects of illumination changes. In order to reduce the size of parameters, we invert the feature maps obtained by convoluting and concatenate them with the original. Besides, we pre-train the proposed network on Places dataset to solve the problem of the scene being unlabeled and difficult to train. Final, the feature vectors are normalized and the similarity of them is calculated by Euclidean distance. Experiments are proved on the CityCenter and the NewCollege dataset [7] to test the proposed approach.

The paper is organized as follows: In Sect. 2, the literature most related to loop closure detection for visual SLAM is reviewed. In Sect. 3, our proposed approach is introduced. Experimental results together with relevant discussion are shown in Sect. 4. Finally, we give a brief conclusion of this paper.

2   Related Work

In this section, we will introduce the previous work about loop closure detection using the traditional methods, which apply handcrafted features. Then, the approaches which use deep learning in the visual loop closure detection will be introduced.

At present, for loop closure detection, the most algorithms match the current position image with the previous position image. In this case, the loop closure detection can be seen as image matching problem. Most algorithms for image matching use handcrafted feature models, such as Bag-of-Words (BoW) [2], Fisher vector (FV) [8] and so on. The BoW that is originally used in document classification is usually applied on loop closure detection for visual SLAM [2, 9]. In visual SLAM, the BoW model extracts a few of visual features from the input image. These visual features are local features (e.g. SIFT [3], SURF [4], ORB [5]) of the images, also are called as visual words in BoW model. And each visual word represents a class of similar features. The BoW model clusters these visual feature descriptors in the extracted image and builds a word dictionary. When a given image arises, it extracts the features from this image and looks in the dictionary for visual words that correspond to the features. The image is depicted as a list of the number of occurrences of every visual word. An example is Fast Appearance-Based Mapping (FAB-MAP) [7], which has successfully introduced BoW in loop closure detection and has a good performance in terms of accuracy and effectiveness. It has already become one of the standard loop closure detection algorithms. Another example is Fisher vector (FV) [8], which establishes a visual word dictionary by adopting a Gaussian mixture model (GMM). It can obtain more useful information from the image than BoW. Zhang et al. [10] applies the binary feature of the image to the incremental loop closure detection, which improves the efficiency of loop closure detection. Different from local features, Liu et al. [11] uses the image's GIST [12] descriptor to extract the global features of the image, which is welcomed in the recent visual SLAM study.

Although these traditional methods have been successful for visual loop closure detection, they have the weakness of lack of robustness with respect to illumination and scale changes. Because in order to obtain the required features, human's expertise and insights dominate the development process. In the scene where the illumination changes obviously, these methods ignore some useful information and the precision of the loop closure detection is reduced.

Very recently, with the development and success of deep learning techniques such as convolutional neural networks (CNN) [6] in image recognition, it is a trend for loop closure detection of visual SLAM that deep learning has been applied to solve this problem. Since 2015, researchers have begun to apply deep learning to visual loop closure detection. Gao et al. [13, 14] proposed a good method that used a modified stacked denoising auto-encoder (SDA) to extract image features and detected loops in the similarity or difference matrix. The SDA is a deep neural network trained in an unsupervised way. Xia et al. [15] applied the cascaded deep learning model PCANet [16] to extract image features into loop closure detection, which is superior to traditional handcrafted features. Hou et al. [17] used position convolutional neural networks (PlaceCNN) to extract image features for loop closure detection in illumination variant environment. Although the algorithm improves the influence of illumination changes on loop closure detection, the extracted feature dimensions are high, which is difficult to meet the real-time requirements of loop closure detection. He et al. [18] put forward a fast and lightweight convolutional neural network (FLCNN) to extract the image features and calculate the similarity matrix. Although the algorithm has improved real-time performance, its precision is reduced. We propose a novel method to improve accuracy based on real-time performance. We perform superpixel processing on the original image. The feature maps obtained by convoluting are inverted and concatenate them with the original. We show that the performance of the approach is better than other methods.

3   Method

In this section, we will show the proposed loop closure detection approach in details.

As described in Fig. 1, the goal is to make full use of image information without being affected by changes in illumination, and to improve the precision of loop closure detection. Different from the traditional approaches, our method uses the simplified CNN model to extract image features. In the image preprocessing, we perform superpixel processing on the original image and the processed image is grayed into the model. The model is pre-trained in the dataset [19], and the pre-trained model acts as a feature extractor. The feature vector of a given image is compared to the previous image feature vectors for similarity. When the similarity surpasses a threshold, a closed loop is considered to occur.

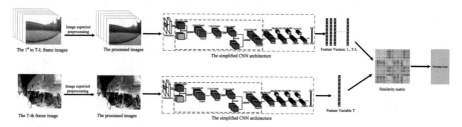

Fig. 1. A framework of the proposed SCNN-based loop closure detection.

3.1   Image Preprocessing

In order to make full use of the image information and reduce the effects of illumination changes, the original image is subjected to superpixel preprocess and then grayscale in the preprocessing. The image is treated as a weighted map by superpixel preprocessing algorithm. Each pixel in image is considered to be the vertices of the graph, and the edges between the two fixed points reflect the difference between two pixels. In addition, color differences between pixels are also calculated and incorporated into the graph according to adaptive thresholds. First, it is necessary for each pixel in the graph to calculate the dissimilarity of its four neighborhoods. Then, the obtained dissimilarity values are arranged in ascending order, and the smallest dissimilarity is selected. Final, the edges with the smallest difference of the edges are combined. If the two vertices of the edge, that is two pixels in the figure, satisfy the following two conditions: (1) they do not belong to one region; (2) the difference between two pixels is smaller than the degree of dissimilarity between the regions which the two pixels belong to; then the threshold is updated and category labels are executed. Otherwise, these three steps are repeated until the desired side is traversed.

3.2 The SCNN Architecture

To reduce the amount of manual operation, a deep learning network is adopted to extract image features instead of handcrafted features for loop closure detection. And the network is further simplified to meet real-time requirements. The architecture of our proposed simplified convolutional neural network (SCNN) is shown in the dark red dotted box in Fig. 1. We reduce the size of the parameters mainly in the black border in the dark red dotted box. We first use 32 convolution kernels to obtain 32 feature maps, then invert these feature maps and concatenate them with the original feature maps to obtain 64 feature maps as Conv1. The same is true of the Conv2. This halved the parameters of the convolution kernel and reduced the model parameters. The Rectified Linear Unit (ReLU) is used as nonlinear activation function both in the convolutional layer and the fully connected layer. The two pooling layers adopt the maximum pooling, which is connected behind the convolutional layer Conv1 and Conv2 to reduce the size of the feature map. In general, there are two parts in the Convx (x = 3, 4, 5), Convx_1 and Convx_2, which all adopt a deep residual [22] module and consist of three convolutional layers with convolution kernel size 1×1, 3×3 and 1×1. And the output of Fc6 is used as the feature vector of the image. The SCNN model is pre-trained by the Places datasets [19]. In order to speed up the training and increase the classification effect during pre-training, batch normalization (BN) [20] is employed after the convolutional layer and the fully connected layer. We adopt the softmax classifier for classification.

3.3 The Similarity Calculation

The most essential part of loop closure detection is to calculate the similarities of two adjacent frames. From this purpose, we extract the features of the image through the SCNN model introduced above, and choose the output of the fully connected layer (Fc6) as the feature vector of the image. We first set up a series of images for similarity comparison. Specifically speaking, as shown in Fig. 1, the images that are compared to the T-th frame for similarity are from frame 1 to frame T-L, where L is the number of frames adjacent to the T-th frame image. Then the extracted feature vector T is compared with the feature vector (1, 2, …, T-L) of the previously extracted images. In addition, the feature vectors are normalized and the similarity of them is calculated by Euclidean distance. The calculation formula is as follow:

$$D(x^i, x^j) = \left\| \frac{V_f^{x^i}}{\left\| V_f^{x^i} \right\|_2} - \frac{V_f^{x^j}}{\left\| V_f^{x^j} \right\|_2} \right\|_2 \tag{1}$$

Where $D(x^i, x^j)$ is the distance between image x^i and x^j, and $V_f^{x^i}$, $V_f^{x^j}$ represent the feature vectors of the fully connected layers extracted from x^i and x^j in the SCNN model, respectively, and $\|\bullet\|_2$ is the L_2-norm of the vector. We use the following formula to define the similarity.

$$S(x^i, x^j) = 1 - \frac{D(x^i, x^j)}{\max\{D(x^i, x^j)\}} \qquad (2)$$

Where $S(x^i, x^j)$ is the degree of similarity between the two feature vectors. To facilitate the calculation, we use the normalized distance to normalize the resulting similarity score to [0, 1]. If the degree of similarity is greater than or equal to a specific threshold, it will be treated as a closed loop.

4 Experimental Result and Evaluation

In this section, we will use two loop closure datasets to test the performance of our proposed methods, including New College and City Center [7]. The two datasets have been widely used to evaluate the loop closure detection algorithm of visual SLAM. There are 1073 and 1237 pairs of images of size 640 × 480 respectively, taken by a mobile robot with two cameras on the left and right side every 1.5 m while traveling through the environment. The ground truth of true loop closures in the datasets is available. For evaluation, we use two indicators of precision-recall (PR) curve [23] and time performance. With the encouragement of [21], we set the L values of the New College and City Center datasets to 100 and 800, respectively. Note that all the experiments are implemented with NVIDIA GTX 980 GPU with 6-GB memory under CUDNN V6 and CUDA 8.0. The software platform used is Ubuntu system. On the Ubuntu system, the TensorFlow framework is used to build the SCNN model.

4.1 The Performance of Loop Detection Algorithm

The difficulty of loop closure detection is whether the loop closure is correctly detected. In order to evaluate our proposed method, we adopt precision-recall (PR) curve. Its calculation formula is as follows:

$$P = \frac{TP}{TP + FP} \qquad (3)$$

$$R = \frac{TP}{TP + FN} \qquad (4)$$

Where P, R represent precision, recall rate respectively. The true positive (TP) is the number of the correct loops detected. The false positive (FP) is the number of the loops that are incorrect by detecting. The false negative (FN) denotes the ground-truth loops which are wrongly detected by the algorithm.

We experimented according to the proposed method in Sect. 3. In order to verify the validity of our proposed approach, we compare it with the other six loop closure detection algorithms in the PR curve. Figure 2 and Fig. 3 respectively show the PR curves of our experimental results using the standard evaluation method in the New College dataset and the City Center dataset. In the charts, we have three PR curves representing the experimental results for all images, left image and right images of the

New College dataset and the City Centre dataset. From the results, we can find that all methods have higher precision when the recall rate is small. As the recall rate increases, the precision of the various methods is gradually decreasing. We also discover that when the recall rate is the largest, the minimum precision of the proposed method is higher than other methods. From these experimental results, we can conclude that our proposed method can obtain better performance in loop closure detection for visual SLAM than other methods.

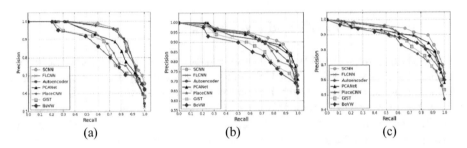

Fig. 2. The experimental results on New College dataset. (a), (b), and (c) show performance comparisons of all images, left images, and right images, respectively.

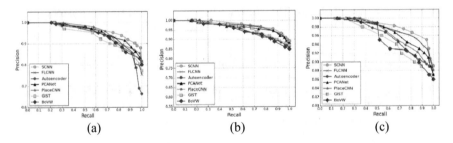

Fig. 3. The experimental results on City Centre dataset. (a), (b), and (c) show performance comparisons of all images, left images, and right images, respectively.

4.2   Computational Time

In order to verify the real-time nature of the method in this paper, we introduced the time spent evaluating the proposed method. We compare the time required by our proposed loop closure detection method with two other loop closure detection algorithms for handcrafted features and four other loop closure detection algorithms based on deep learning. The average cost time of each image using different algorithm is shown in Table 1. The time mentioned here is the average of 1000 images, which only contains the time of feature extraction. As can be seen from Table 1, the time taken by our proposed method is 0.017, which is an order of magnitude faster than the two methods of handcrafted features. Compared with the other four deep learning methods, our method takes less time than the other three methods, but it takes a little longer than

FLCNN [17]. Because FLCNN is a specially designed fast convolutional neural network model, real-time performance is better than our method. In general, our approach can satisfy the real-time requirements for loop closure detection in visual SLAM.

Table 1. The average cost time of each image for different algorithm

Different algorithm	Time (s)
BoVW [2]	0.567
GIST [11]	1.787
Autoencoder [13]	0.020
PCANet [15]	0.024
PlaceCNN [17]	0.052
FLCNN [18]	0.013
Our Approach	**0.017**

5 Conclusion

In this paper, we propose a loop closure detection method for visual SLAM based on simplified convolutional neural networks. Compared to traditional handcrafted features, CNN features can learn more useful information from a given image for loop closure detection. Therefore, our proposed simplified CNN network model is trained by the Places dataset and then is used as a feature extractor. Before extracting features, the original image is first processed with superpixel preprocess in order to get more useful information from the image without respect to illumination and scale changes. Then we use the Euclidean distance to compare the extracted feature vectors to determine if there is a loop closure. The experimental results in Sect. 4 show that compared with other traditional methods, the SCNN-based method is feasible for loop closure detection and has good performance in both precision-recall rate and real-time. Our future work is to increase the constraints to avoid false loop closure detection and speed up the comparison of feature vector comparisons to reduce the time for loop closure detection.

References

1. Konolige, K., Agrawal, M.: FrameSLAM: from bundle adjustment to real-time visual mapping. IEEE Trans. Rob. **24**(5), 1066–1077 (2008)
2. Filliat, D.: A visual bag of words method for interactive qualitative localization and mapping. In: IEEE International Conference on Robotics and Automation, pp. 3921–3926. IEEE, Roma (2007)
3. Lowe, D.G.: Distinctive image features from scale-invariant keypoints. Int. J. Comput. Vision **60**(2), 91–110 (2004)
4. Bay, H., Tuytelaars, T., Gool, L.: SURF: speeded up robust features BT - computer vision–ECCV 2006. In: Computer Vision-ECCV, vol. 3951, pp. 404–417 (2006)
5. Rublee, E., Rabaud, V., Konolige, K.: Orb: an efficient alternative to sift or surf. In: IEEE International Conference on Computer Vision, pp. 2564–2571. IEEE, Barcelona (2006)

6. Bengio, Y., Courville, A., Vincent, P.: Representation learning: a review and new perspetives. IEEE Trans. Pattern Anal. Mach. Intell. **35**(8), 1798–1828 (2012)
7. Cummins, M., Newman, P.: FAB-MAP: probabilistic localization and mapping in the space of appearance. Int. J. Robot. Res. **27**, 647–665 (2008)
8. Daniilidis, K., Maragos, P., Paragios, N.: Improving the fisher kernel for large-scale image classification. In: ECCV, pp. 143–156 (2010). https://doi.org/10.1007/978-3-642-15561-1_11
9. Mur-Artal, R., Tardos, J.D.: ORB-SLAM2: an open-source SLAM system for monocular, stereo, and RGB-D cameras. In: IEEE Transactions on Robotics, pp. 1–8 (2017)
10. Zhang, G., Lilly, M.J., Vela, P.A.: Learning binary features online from motion dynamics for incremental loop-closure detection and place recognition, pp. 765–772 (2016)
11. Liu, Y., Zhang, H.: Visual loop closure detection with a compact image descriptor. In: IEEE/RSJ International Conference on Intelligent Robots and Systems, pp. 1051–1056. IEEE, Vilamoura-Algarve (2012)
12. Oliva, A., Torralba, A.: Modeling the shape of the scene: a holistic representation of the spatial envelope. Int. J. Comput. Vision **42**(3), 145–175 (2001)
13. Gao, X., Zhang, T.: Loop closure detection for visual SLAM systems using deep neural networks. In: 34th Chinese Control Conference (CCC). IEEE (2015)
14. Gao, X., Zhang, T.: Unsupervised learning to detect loops using deep neural networks for visual SLAM system. Autonomous Robots **41**(1), 1–18 (2017)
15. Xia, Y., Li, J., Qi, L., et al.: Loop closure detection for visual SLAM using PCANet features. In: International Joint Conference on Neural Networks, pp. 2274–2281. IEEE (2016)
16. Chan, T.H., Jia, K., Gao, S., et al.: PCANet: a simple deep learning baseline for image classification. IEEE Trans. Image Process. **24**(12), 5017–5032 (2015)
17. Hou, Y., Zhang, H., Zhou, S.: Convolutional neural network based image representation for visual loop closure detection. In: IEEE International Conference on Information and Automation, Lijiang, August 2015, pp. 2238–2245 (2015)
18. He, L., Chen, J.T., Zeng, B.: A fast loop closure detection method based on lightweight convolutional neural network. Comput. Eng. **44**(6), 182–187 (2018)
19. Zhou, B., Lapedriza, A., Khosla, A., et al.: Places: a 10 million image database for scene recognition. IEEE Trans. Pattern Anal. Mach. Intell. (2017)
20. Ioffe, S., Szegedy, C.: Batch normalization: accelerating deep network training by reducing internal covariate shift, pp. 448–456 (2015)
21. Bai, D., Wang, C., Zhang, B., et al.: Matching-range-constrained real-time loop closure detection with CNNs features. Rob. Biomimetics **3**(1), 15–21 (2016)
22. He, K., Zhang, X., Ren, S., et al.: Deep residual learning for image recognition (2015)
23. Xia, Y., Li, J., Qi, L., et al.: An evaluation of deep learning in loop closure detection for visual SLAM. In: IEEE International Conference on Internet of Things. IEEE (2018)

Development of Optimal Pump Operation Method for Urban Drainage Systems

Yoon Kwon Hwang[1], Soon Ho Kwon[1], Eui Hoon Lee[2],
and Joong Hoon Kim[3(✉)]

[1] Department of Civil, Environmental and Architectural Engineering,
Korea University, Seoul, Korea
[2] School of Civil Engineering, Chungbuk National University, Cheongju, Korea
[3] School of Civil, Environmental and Architectural Engineering,
Korea University, Seoul, Korea
jaykim@korea.ac.kr

Abstract. Recently, climate change and urbanization have been caused by urban flood. To mitigate flood, there are structural measures such as detention reservoir and pump station and non-structural measures such as pump operation rules and gate operation rules. However, structural measures cannot cope with rainfall stronger than design rainfall intensity and need to construct to develop flood mitigation capacity. Urban drainage system needs non-structural measures that maximize the efficiency of structural measures. The efficient operation of the pump as one of non-structural measures can develop the discharge capacity of urban drainage system to prevent urban flood. In this study, we proposed advanced optimal pump operation at Mokgam stream located Seoul and Gyeonggi-do. Advanced optimal pump operation is considering the water level of the vulnerable points and the water level of reservoirs. Risk of flooding is predicted by the water level of vulnerable point and reservoir. Therefore, pump station is able to proactively start the pump compared to the current pump operation, reservoir secures free space to develop flood mitigation effect and discharge capacity.

Keywords: Optimal pump operation · Harmony search algorithm ·
Multiobjective optimization · Non-structural measures

1 Introduction

Flood mitigation measures are classified into two approaches: structural measures and non-structural measures. Structural measures are the installation of drainage facilities such as pump stations and detention reservoirs. Structural measures have the advantage of reducing flooding directly by using drainage facilities. However, structural measures need a lot of spaces and costs to increase capacity of flood mitigation. Non-structural measures mean flood mitigation measures that do not pass through structures. Non-structural measures can improve flood mitigation capacity at without a lot of spaces and costs. Therefore, non-structural measures become more important part of flood

J. H. Kim et al. (Eds.): ICHSA 2019, AISC 1063, pp. 63–69, 2020.
https://doi.org/10.1007/978-3-030-31967-0_7

mitigation measures than before. The optimal pump operation can improve drainage facilities' capacity as non-structural measures.

Early studies of optimal pump operation considered minimizing the index related to flooding without maintenance evaluation (Song et al. 2014; Lee et al. 2016). The total flooding volumes indicate the capacity of drainage facilities indirectly. However, these operations do not consider continuous maintenance of pump station. To overcome this shortcoming, later studies considered the total flooding volumes and several considerations simultaneously that represent maintenance of pump station (Jafari et al. 2018).

This study suggests advanced pump operation considering both flooding and operation cost and optimum on/off level of pump station. Multi-objective Harmony Search Algorithm (MOHS) is used as technique of optimization. This advanced optimal pump operation has a beneficial effect on comprehensive economic implications, including the flood damage and the maintenance costs of the pump.

2   Methodology

2.1   Advanced Pump Operation

The water level of reservoir (h_r) linked with pump station is generally standard for pump operation. However, since this pump operation cannot predict the inflow volumes, there is high probability of flooding if sudden influx occurs. To overcome this drawback, the water level of vulnerable points (h_n) are additional standard of pump operation. Vulnerable points mean where the first flood occurs or where the maximum flooding volumes occurs. In advanced operation, if one of the standards that RL_{on} and the start up level of vulnerable point (NL_{on}) is met, pump operates to drain water of reservoir. If all of the standards that RL_{off} and the shut off level of vulnerable point (NL_{off}) are met, pump shuts off. This operate method improves the overall discharge capacity of urban drainage system and reduces the risk of flooding in the upstream watershed. Figure 1 shows a schematic diagram of current pump operation. Figure 2 shows a schematic diagram of advanced pump operation.

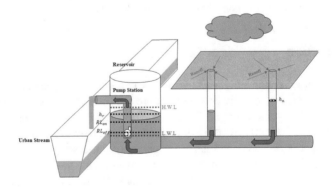

Fig. 1. Schematic diagram of current pump operation

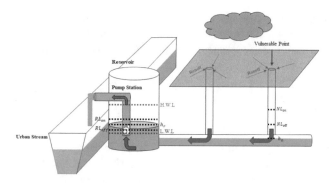

Fig. 2. Schematic diagram of advanced pump operation

2.2   Multi-objective Harmony Search Algorithm

Geem et al. (2001) proposed Harmony search algorithm (HSA) as metaheuristic optimization algorithm. Authors got inspired by improvisation of jazz musicians. When jazz musicians play improvisation, they adjust pitch of sound or choose the best harmony that experienced before. When this phenomenon is applied to optimization, the sounds of each jazz musicians become decision variables, and the harmony is considered the objective function.

A multi-objective harmony search (MOHS) is developed HSA that including a non-dominated sorting method and the crowding distance concept (Deb et al. 2002). Trade-off means a conflicting relationship about several objective functions. Two objective functions in trade-off cannot find the optimal solution through optimization. Instead, the Pareto-optimal solutions can be obtained using MOHS.

2.3   Optimization Problem

In this study, risk of flooding and operation cost of pump are considered as objective functions. The risk of flooding is determined through total flooding volumes. The per minute flooding volumes are calculated by the continuity equation. The total flooding volumes are the first objective function and shown below:

$$Minimize\ F1 = \sum_{m=1}^{N} \left(\sum_{y=1}^{Nm} Inflow_y - Outflow_y - Extra\ storage_y \right)_m \quad (1)$$

where, F_1 is the total flooding volumes; N means the total minutes of simulation time; and Nm is the number of Nodes including reservoirs linked with pump station.

When Eq. (1) is calculated, if per minute inflow volumes is bigger than the sum of the extra storage and per minute outflow volumes, flooding is the difference between the per minute inflow and sum of the extra storage and per minute outflow volumes. If inflow volumes are smaller than sum of the extra storage and per minute outflow volumes, per minute flooding volume is $0\,m^3$.

The total operation costs are the second objective function and are calculated by power usage equation and conversion constant of Korea Electronic Power Corporation (KEPC). The equation used to calculate total operation cost of pump is shown below:

$$\text{Minimize } F_2 = \sum_{t=1}^{N} \left(\left(0.163 \times \frac{r \times Q \times H}{\vartheta} \right) \times (1+\alpha)(\text{kW}) \times 0.0016 \left(\frac{\text{won}}{\text{kW}} \right) \right)_t$$

(2)

where, F_2 is the total pump operation cost; r is the density of water; Q is the pump outflow; H is the total pump head; ϑ is the pump efficiency; and α the excess rate of motors. The density of water, the total pump head, the pump efficiency, and the excess rate of motors are constant.

subject to :

$$Q_i(x) \begin{cases} 0, & \text{if } h_{r,i} < RL_{off,i} \text{ and } h_{n,i} < NL_{off,i} \\ (0, Q_{pump,i}), & \text{if } RL_{off,i} < h_{r,i} < RL_{on,i} \text{ and } NL_{off,i} < h_{n,i} < NL_{on,i} \\ Q_{pump,i}, & \text{if } RL_{on,i} < h_{r,i} \text{ or } NL_{on,i} < h_{n,i} \end{cases}$$

(3)

$i=1$ to 2

where, Q_i is the outflow of Gaebong i pump station same as discharge amounts. Pump discharges are determined by each pump curve rule. If both $h_{r,i}$ and $h_{n,i}$ are smaller than each shut off level $\left(RL_{off,i} \text{ and } NL_{off,i} \right)$ Gaebong i pump station shuts off. If $h_{r,i}$ or $h_{n,i}$ is bigger than each start up level $\left(RL_{on,i} \text{ and } NL_{on,i} \right)$ Gaebong i pump station operates. If both $h_{r,i}$ and $h_{n,i}$ are located between start up level and shut off level, the previous water level determines pump discharges. If the previous water level is bigger than start up level, pump discharges are following pump curve. Otherwise, pump discharges are 0.

$$L.W.L < RL_{on,i} < H.W.L$$

(4)

$$RL_{off,i} \le RL_{on,i} - 0.5 \text{ m}$$

(5)

$$0 < NL_{on,i} < h_{max,i}$$

(6)

$$0 < NL_{off,i} < NL_{on,i}$$

(7)

where, $RL_{on,1}$ and $RL_{on,2}$ are the start up level as reservoirs; $NL_{on,1}$ and $NL_{on,2}$ are the start up level as vulnerable point; $RL_{off,1}$ and $RL_{off,2}$ are the shut off level as reservoirs; and $NL_{off,1}$ and $NL_{off,2}$ are the shut off level as vulnerable point. Therefore, $RL_{on,1}$ and $RL_{on,2}$ must be in range between low water level and high water level. To prevent cavitation, $RL_{off,1}$ and $RL_{off,2}$ must be 0.5 m smaller than each start up level. $NL_{on,1}$ and $NL_{on,2}$ must be in range between 0 m and Maximum depth of vulnerable point. $NL_{off,1}$ and $NL_{off,2}$ must be smaller than each start up level.

3   Study Area

Study Area is Mokgam stream watershed located between Seoul and Gyunggi-do. Mokgam stream is upstream of Anyang stream that a tributary of Han river. Mokgam stream is urban stream that can cause risky urban flooding. The sewer network of Mokgam stream watershed expressed in the form of EPA-SWMM is shown in Fig. 3. Table 1 shows information of two pump station belong to drainage stream of Mokgam stream watershed.

Fig. 3. Mokgam watershed on EPA-SWMM

Table 1. Pump station list at Mokgam stream watershed

Status	Gaebong 1 pump station	Gaebong 2 pump station
High water level (m)	10.9	8.0
Low water level (m)	6.0	3.0
Pump start up level (m)	6.5	4.0
Pump capacity (m³/min)	9,440	4,800

4   Application Results

Rainfall event occurred at 23 September 2010, historical flooding occurred in Mokgam stream watershed, was used input data to simulate extreme rainfall that causes an urban flood. In current pump operation, The total flooding volumes are 104,668.2 m³. Total pump operation costs are 2,025,615 won. The per minute flooding volumes are shown in Fig. 4. Node N-17-05425 is the first flooding place and Node N-17-00425 is the maximum flooding place as vulnerable nodes.

In this study, the MOHS parameters were set as follows: harmony memory size (HMS) = 10, harmony memory considers rates (HMCR) = 0.8, pitch adjust rates (PAR) = 0.2, bandwidth (BW) = 0.01, and the maximum number of iterations = 50,000.

In a result of MOHS, Total flooding volumes are located between 6,833.4 m³ and 1,266,000 m³. Total operation cost are located 1,67,469 won and 1,945,755 won. Compared with the current pump operation, total flooding volumes are reduced by

93.47% or increased by 1,109% and total operation cost are reduced by 3.94% to 91.73%. Result of current pump operation is shown in Fig. 5 as an orange dot. Pareto-optimal solutions are shown in Fig. 5 as blue dots.

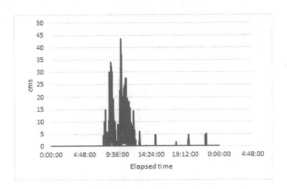

Fig. 4. Per minute flooding volumes in current pump operation

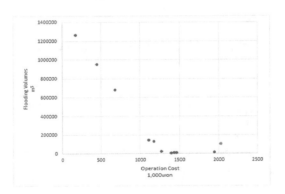

Fig. 5. Pareto-optimal solutions

Some Pareto-optimal solutions suggest advanced optimal pump operation have smaller flooding volumes and smaller operation cost. Of the values with fewer total flooding volumes than the one currently operating, Pareto optimal solution with the lowest total flooding volumes is a reasonable solution and shown in Table 2. Compared with the current pump operation, Total flooding volumes are decreased by 93.47% and Total pump operation costs are decreased by 31.34%. The per minute flooding volumes both the current pump operation and the optimal pump operation are shown in Fig. 6. Therefore, this optimal pump operation is more reasonable operation than the current pump operation.

Table 2. Optimum on/off level from Pareto-optimal solutions (m)

$RL_{on,1}$	$RL_{on,2}$	$RL_{off,1}$	$RL_{off,2}$	$NL_{on,1}$	$NL_{on,2}$	$NL_{off,1}$	$NL_{off,2}$
7.54	4.99	6.64	4.44	1.24	0.55	0.11	0.44

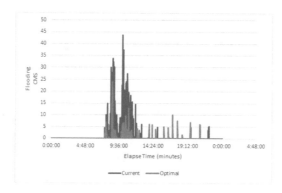

Fig. 6. Per minute flooding volumes

5 Conclusions

This study demonstrates advanced pump operation considering both the pump operation cost and the total flooding volumes as a non-structural measure for flood mitigation. Optimum on/off level is shown Table 2. Compared to the current pump operation, Optimal solution obtains lower total flooding volumes and lower pump operation costs. Therefore, advanced pump operation is non-structural measure to increase the efficiency of urban drainage system.

This study has several limitations that future research must address. First, advanced pump operation should be applied to various rainfalls and areas. The optimum on/off criterion for various design frequency will be guideline for each rainfall and each area. Second, resilience is more proper index than total flooding volumes to evaluate flood mitigation capacity of urban drainage system.

Acknowledgements. This work was supported by a grant from The National Research Foundation (NRF) of Korea, funded by the Korean government (MSIP) (No. 2016R1A2A1A050-05306).

References

Deb, K., Pratap, A., Agarwal, S., Meyarivan, T.: A fast and elitist multi-objective genetic algorithm: NSGA-II. IEEE Trans. Evol. Comput. **6**(2), 182–197 (2002)

Geem, Z.W., Kim, J.H., Loganathan, G.V.: A new heuristic optimization algorithm: harmony search. Simulation **76**(2), 60–68 (2001)

Jafari, F., Mousavi, S.J., Yazdi, J., Kim, J.H.: Real-time operation of pumping systems for urban flood mitigation: single-period vs. multi-period optimization. Water Resour. Manage. **32**(14), 4643–4660 (2018)

Lee, E.H., Lee, Y.S., Joo, J.G., Jung, D., Kim, J.H.: Flood reduction in urban drainage systems: cooperative operation of centralized and decentralized reservoirs. Water **8**(10), 469 (2016)

Song, Y.H., Park, M.J., Lee, J.H.: Analysis of urban inundation reduction effect by early operation of drainage pumping station. J. Korean Soc. Hazard Mitig. **14**(2), 267–276 (2014)

Parameter Estimation of Storm Water Management Model with Sewer Level Data in Urban Watershed

Oseong Lim[1] , Young Hwan Choi[1] , Do Guen Yoo[2] ,
and Joong Hoon Kim[1(✉)]

[1] Korea University, Seoul 02841, Republic of Korea
Jaykim@korea.ac.kr
[2] The University of Suwon, Hwaseong 18323, Republic of Korea

Abstract. The rainfall-runoff analysis model in urban watersheds should be constructed to establish flood damage countermeasures. The SWMM (Storm Water Management Model) is a representative model for rainfall-runoff analysis of urban watersheds. While this model is based on many parameters and provides relatively reliable results, it contains many ambiguous parameters. Therefore, parameter estimation is essential for rainfall-runoff analysis model and can be done using optimization algorithms. Harmony search algorithm is used to automatically estimate the parameters of the SWMM. Unlike the previous studies, the parameters are estimated by considering not only the inflow data but also the sewer level data. Parameter estimation is applied to the flood simulation on the catchment of Yongdap pump station basin, Seongdong-gu, Seoul, South Korea. The results estimated by supposed model are reliable in terms of both inflow and sewer level. The verification results of the calibrated model show the error within 5%, which are within the allowable error range.

Keywords: SWMM · Parameter estimation · Calibration · Sewer level data

1 Introduction

The growth of severe rain storm in the world has increased flood damage severely, and the precipitation distribution is getting more erratic. The unpredictability in precipitation increases the seriousness of the existing flood damage especially during rainy seasons. Structural measures such as installation or expansion of drainage facilities and improvement of reverse gradient of sewer pipes can be applied to decrease the flood damage. However, these measures require high cost, lots of time, and large site. For these reasons, non-structural measures can be alternatives, and a rainfall-runoff analysis model must be established to apply non-structural measures.

The SWMM (Storm Water Management Model) is a representative model for rainfall-runoff analysis of urban watersheds. While this model is based on many parameters and provides relatively reliable results, it contains many ambiguous parameters. Therefore, parameter estimation is essential and can be done using optimization algorithms.

© Springer Nature Switzerland AG 2020
J. H. Kim et al. (Eds.): ICHSA 2019, AISC 1063, pp. 70–75, 2020.
https://doi.org/10.1007/978-3-030-31967-0_8

In present study, harmony search algorithm, one of the widely known meta-heuristic algorithms is used to automatically estimate the parameters of the SWMM. The parameters are estimated by considering not only the inflow data but also the sewer level data, unlike the previous other studies. The simulation results of the calibrated model are compared with the observed values to verify the accuracy of the estimated parameters. Finally, the calibrated model of the sewer network with parameter estimation is verified with other rainfall events to confirm the validity of the model.

2 Methodology

2.1 Performance Measures

The combined term of sewer level data and inflow is considered as objective function to estimate the parameter of the SWMM. The objective functions used in parameter estimation of the SWMM were examined in previous studies, and finally the modified Nash-Sutcliffe Efficiency (NSE) form is adopted as the objective function as follows:

$$\frac{\sum_{t=1}^{N}\left(d_t^{ob} - d_t^{sm}\right)^2}{\sum_{t=1}^{N}\left(d_t^{ob} - d_{mean}^{ob}\right)^2} + \frac{\sum_{t=1}^{N}\left(q_t^{ob} - q_t^{sm}\right)^2}{\sum_{t=1}^{N}\left(q_t^{ob} - q_{mean}^{ob}\right)^2} \tag{1}$$

where N is the number of time series; d_t^{ob} and d_t^{sm} are the observed and simulated sewer level data (m), respectively, at the tth time period; q_t^{ob} and q_t^{sm} are the observed and simulated inflow data (m^3/s), respectively, at the tth time period; d_{mean}^{ob} and d_{mean}^{sm} are the mean value of the observed sewer level data (m) and observed inflow data (m^3/s). The modified form has a value between 0 and 1, and unlike the original NSE, the closer to 0, the better the result.

It is necessary to construct the evaluation index in order to assess the calibrated model with the objective function. For model assessment, Root Mean Square Error (RMSE) and NSE are implemented as evaluation indices (Table 1).

Table 1. Evaluation indices for model assessment

Evaluation index	Equation	Optimal value
RMSE	$\sqrt{\dfrac{\sum_{t=1}^{N}\left(X_t^{ob} - X_t^{sm}\right)^2}{N}}$	0
NSE	$1 - \dfrac{\sum_{t=1}^{N}\left(X_t^{ob} - X_t^{sm}\right)^2}{\sum_{t=1}^{N}\left(X_t^{ob} - X_{mean}^{ob}\right)^2}$	1

2.2   Model Overview

The SWMM, one of the rainfall-runoff models, can make various simulations on the runoff caused by rainfall events in urban watershed. There are several modules in the SWMM, among which modules on the runoff analysis and flood routing analysis play a key role. In the runoff analysis module, simulations such as evaporation, infiltration and pollutant transport are performed in the watershed. In the flood routing analysis module, simulation of flood routing is performed in the conduit, waterway, reservoir and pump.

In order to estimate the parameters of the SWMM, the harmony search algorithm is applied as an optimization algorithm. The harmony search algorithm which is a meta-heuristic algorithm that focuses on combining chords when playing instruments was suggested by Geem et al. (2001). The harmony search algorithm and the SWMM are connected to configure the parameter estimation model.

2.3   Application Network

The parameter estimation method was generated from Yongdap pump station basin, Seongdong-gu, Seoul, South Korea. The watershed area is 34.7 ha, and the land use is 80% for residential area, 6% for public area and 14% for road and pump station area. The area is a typical urban watershed with urbanization. The Yongdap pump station located at the outfall of watershed discharges rainwater from the reservoir to the Cheonggyecheon River located on the left side of the watershed basin, and has 12,100 m^3 of reservoir volume and 352 m^3/min of pumping capacity.

2.4   Estimated Parameters

In SWMM, the parameters can be divided into physical and hydrologic parameters. Physical parameters can be defined as specific values, such as an area of a basin or an extension of a conduit. On the other hand, hydrologic parameters are needed to estimate, because accurate values cannot be specified.

Table 2. Types and ranges of estimated parameters

Estimated parameters	Lower bound	Upper bound
Percent of impervious area (%)	65	99
Manning's N for pervious area	0.15	0.40
Manning's N for impervious area	0.010	0.015
Depth of depression storage on pervious area (mm)	2.5	5.0
Depth of depression storage on impervious area (mm)	1.3	2.5
Percent of impervious area with no depression storage (%)	10	30
Characteristic width of overland flow path (m)	0.7	1.3
Average surface slope (%)	0.7	1.3

For this reason, hydrologic parameters are selected as the estimated parameters. Especially, parameters of the subcatchment rather than the conduit are selected as the estimated parameters, because the parameters of the conduit do not have much influence compare to the parameters of the subcatchment. Therefore, hydrologic parameters of subcatchment are estimated and a total of eight parameters are selected with reference to previous studies (Table 2).

The last two parameters are unique values of each subcatchment that estimated by the geographic information system such as GIS or RS, and it is difficult to arbitrarily set the range. In optimization process, these two parameters are estimated by multiplying the existing values with a certain range.

3 Application Results

3.1 Model Calibration

The parameter estimation of the SWMM in application network watershed is performed by applying suggested optimization method. The rainfall applied to the model calibration is the rainfall event occurred on July 13, 2013. The total daily rainfall is 175 mm and the hourly maximum rainfall is 42.5 mm/h. In the optimization process, HM, HMCR, and PAR are set to 30, 0.95, and 0.7, respectively.

The optimization results of parameters estimation are shown in Table 3. These results indicate the degree of error in inflow and sewer level when simulating the sewer network model with parameters estimated as an objective function. In the derived evaluation index, since the NSE is non-dimensional, it does not show a large difference between the value based on the inflow and the value based on the sewer level. However, in the case of RMSE, the value calculated based on the sewer level is much smaller than based on the inflow. This is because RMSE has different units at (m^3/s) and (m) at inflow and sewer level, respectively.

Though some errors occur in the results, most of them are similar to the observed values. In addition, due to the characteristics of the inflow and the sewer level which show similar behavior, there is no significant difference between the evaluation index of inflow and sewer level. Compared with the previous study of NSE confidence level of 0.8 in the rainfall-runoff model and other studies related to parameter calibration (Tan et al. 2008; Mancipe-Munoz et al. 2014; Kang et al. 2012), the NSE of 0.91 or more as a result of the calibration is considered to be a reliable result.

Table 3. Assessment results for the model calibration

Classification	Inflow	Sewer level
RMSE	0.241	0.047
NSE	0.942	0.932

3.2 Model Verification

Since the SWMM has contains many parameters, the reliability of the model can vary depending on the input rainfall. For example, two models can be constructed with different combinations of parameters. Although the results of both models show almost similar results for one rainfall event, they can show significant differences for other rainfall events. Therefore, a model that has been calibrated should be verified by applying different rainfall events.

The calibrated model is verified by applying two different rainfall events, occurred on July 22, 2013 and July 24, 2014. Each of the observed data and the parameters of the SWMM is calculated in the same manner as used for calibration of the model.

In the case of two kinds of rainfall events, the tendency of sewer level change with time is very similar to the tendency of inflow change with time. In the evaluation indices, a larger error than the calibration result occurred, but it can be considered that it is within the allowable error range because the error range is within 5%. In addition, NSE is 0.86 or more, which is a reliable result of actual rainfall-runoff analysis for both rainfall events (Table 4).

Table 4. Assessment results for the model verification

Classification	Rainfall events	Inflow	Sewer level
RMSE	July 22, 2013	0.362	0.066
	July 24, 2014	0.133	0.034
NSE	July 22, 2013	0.903	0.876
	July 24, 2014	0.889	0.868

4 Conclusions

In this study, parameter estimation of SWMM was performed to analyze rainfall-runoff in urban watershed using the harmony search algorithm, which is one of the meta-heuristic algorithm. The study network for estimating parameters of SWMM is the Yongdap pump station basin located in Seongdong-gu, Seoul, South Korea. The inflow and the sewer level used for building the input data were estimated using pump station records and sewer level monitoring records, respectively. The total number of parameters estimated in the study is 8, which was changed within a reasonable range.

The parameters were estimated by taking into account not only the inflow used in the existing method but also the sewer level together with the objective function. In order to apply the optimization problem, both of inflow and sewer level were set as the objective function and the parameters estimation of the SWMM was performed. In the final calibrated model, RMSE and NSE were 0.241 and 0.942, respectively, based on inflow and 0.047 and 0.932, respectively, based on the sewer level. It seems that reasonable results had been obtained in both the inflow and the sewer level. In particular, considering the inflow is estimated based on the operation records of the pump station, it can be considered that the parameters are estimated within an allowable error considering that the error of the measured value is relatively large.

In addition, the final calibrated model was verified for two different rainfall events, and a reasonable result was obtained with a reasonable range of error. It is considered that the parameters estimation using the sewer level data of the SWMM, which is a rainfall-runoff model, is properly performed. The verified model can be considered to reflect the characteristics of the watershed, and can be used for rainfall-runoff analysis and prediction of inflow, etc.

Acknowledgements. This work was supported by a grant from The National Research Foundation (NRF) of Korea, funded by the Korean government (MSIP) (No. 2016R1A2A1A-05005306).

References

Cho, J.H., Seo, H.J.: Parameter optimization of SWMM for runoff quantity and quality calculation in a eutrophic lake watershed using a genetic algorithm. Water Sci. Technol. Water Supply **7**(5–6), 35–41 (2007)

Di Pierro, F., Khu, S.T., Savić, D.: From single-objective to multiple-objective multiple-rainfall events automatic calibration of urban storm water runoff models using genetic algorithms. Water Sci. Technol. **54**(6–7), 57–64 (2006)

Kang, T.U., Lee, S.H., Kang, S.U., Park, J.P.: A study for an automatic calibration of urban runoff model by the SCE-UA. J. Korea Water Resour. Assoc. **45**(1), 15–27 (2012)

Krebs, G., Kokkonen, T., Valtanen, M., Koivusalo, H., Setälä, H.: A high resolution application of a stormwater management model (SWMM) using genetic parameter optimization. Urban Water J. **10**(6), 394–410 (2013)

Lee, J.M., Jin, K.N., Kim, Y.J., Yoon, J.R.: A study on the Establishment of Reasonable Guidelines for Prior Review System on the Influence of Disasters (2010)

Liong, S.Y., Chan, W.T., Lum, L.H.: Knowledge-based system for SWMM runoff component calibration. J. Water Resour. Plan. Manage. **117**(5), 507–524 (1991)

Mancipe-Munoz, N.A., Buchberger, S.G., Suidan, M.T., Lu, T.: Calibration of rainfall-runoff model in urban watersheds for stormwater management assessment. J. Water Resour. Plan. Manage. **140**(6), 05014001 (2014)

Tan, S.B., Chua, L.H., Shuy, E.B., Lo, E.Y.M., Lim, L.W.: Performances of rainfall-runoff models calibrated over single and continuous storm flow events. J. Hydrol. Eng. **13**(7), 597–607 (2008)

Multiobjective Parameter Calibration of a Hydrological Model Using Harmony Search Algorithm

Soon Ho Kwon[1] , Young Hwan Choi[2] , Donghwi Jung[3] ,
and Joong Hoon Kim[4(✉)]

[1] Department of Civil, Environmental and Architectural Engineering,
Korea University, Seoul, South Korea
[2] Research Center for the Disaster and Science Technology, Korea University,
Seoul, South Korea
[3] Department of Civil Engineering, Keimyung University, Daegu, South Korea
[4] School of Civil, Environmental and Architectural Engineering,
Korea University, Seoul, South Korea
Jaykim@korea.ac.kr

Abstract. The tank model among various deterministic Rainfall-Runoff (RR) models is often preferred for its simple concepts by many previous studies. However, it requires much time and effort to obtain better results owing to the need to calibrate a number of parameters in the tank model. The demand for an automatic calibration method has been increasing. The success of heuristic optimization algorithms enables many researchers to focus on the other aspect of the various objective function for tank model rather than parameter calibration. In this study, Multi-objective Harmony Search Algorithm was performed for an automatic calibration. The proposed study enables parameter calibration of four storage types of tank model with 14 parameters and six scenarios based on the four objective functions. A proposed tank model, which determines optimal solution (e.g., parameters of tank model; calibration results) under four different objective functions, respectively, were compared to demonstrate tradeoff relationship between measurements data and observation data for a Daecheong dam watershed in South Korea. The proposed tank model with six different scenarios could be a successful alternative RR model with its increased accuracy and Pareto solution.

Keywords: Multi-objective Harmony Search Algorithm · Tank model · Rainfall-runoff model

1 Introduction

A rainfall–runoff (RR) models are a mathematical model used to describe the RR process of a natural river watershed, which generally produces a surface output (runoff hydrograph) using a input (hyetograph) (Jakeman 1993). During the past two decades, the development and calibration of various RR model have been applied to estimate the parameter of automatic calibration approaches using Genetic Algorithm (Holland 1993)

© Springer Nature Switzerland AG 2020
J. H. Kim et al. (Eds.): ICHSA 2019, AISC 1063, pp. 76–81, 2020.
https://doi.org/10.1007/978-3-030-31967-0_9

and Powell's method. The various RR models have been developed and are classified into lumped and distributed models. The former is based on the assumption that the entire basin has a homogeneous hydrological characteristic, whereas the latter divides the basin into elementary unit areas resembling a grid network to consider the spatial variability of the basin characteristics.

This study introduces various balancing approaches for the exploration and exploitation in the multi-objective automatic calibration for hydrological model (i.e., the modified tank model). The proposed tank model composed parameter calibration of four storage types of tank model with 14 parameters. In addition, the formers proposed the combination of four objective functions (i.e., peak discharge; peak runoff time; total runoff; RMSE) as six scenarios. A proposed tank model, which determines optimal solution (e.g., parameters of tank model; calibration results) based on the 6 scenarios for the four objective functions, respectively, were compared to identify tradeoff relationship between measurements data and observation data for a Daecheong dam watershed in South Korea.

2 Multi-objective Parameter Calibration of a Hydrological Model

2.1 Modified Tank Model

The tank models are intended to simulate either long-term runoff from watersheds of flood events simulated by a combination of vessels. These models are generally classified as deterministic, lumped, linear, continuous and time-invariant models. The non-linear based on model is relaxed through the tank model proposed in this study. The proposed tank model in this study, are composed of a series of vertically laid storages.

For the proposed tank model with four tanks, the outputs through the side outlets of the first (located at the top), second, third and fourth (located at the bottom) tanks are composed as surface runoff, intermediate runoff, sub-base runoff and base flow, respectively. And the output from the bottom outlet of the first tank could be considered as infiltration and the outputs from the bottom outlets of the other tanks could be regarded as percolation. This proposed model is composed of four tanks. The first tank has two side outlets and the other tanks have one side outlet.

2.2 Parameters Calibration Approach

Recently, the automatic calibration procedures have been extended to handle multiple criteria (Gupta et al. 1998; Yapo et al. 1998; Cheng et al. 2002). In general, the criteria most commonly used in the literature are expressed as an objective function, such as the root mean squared error (RMSE) evaluated on either the stream flows or the log of the stream flows, the daily root mean square (DRMS), and so on (Gupta et al. 1998; Boyle et al. 2000). However, the parameters calibration based on the traditional statistical approach have been presented a variety of uncertainties involved by Cheong et al. (2005).

To enhance a parameter calibration approach, formers are introduced criteria for flood forecasting which three conditions to evaluate the parameter calibration performance of a RR models, i.e. three criteria conditions relative to the peak discharge, peak time and total runoff volume among the calibrated and validated historical flood events, respectively. An aforementioned objective functions are used to propose tank model for parameter calibration performance in this study. In addition, we composed the various objective functions as 6 scenarios (Table 1). The various objective functions are expressed as

$$\text{Maximize } R_{peak\ discharge} = \frac{M_{pd}}{n} \times 100\% \tag{1}$$

$$\text{Maximize } R_{peak\ time} = \frac{M_{pt}}{n} \times 100\% \tag{2}$$

$$\text{Maximize } R_{runoff} = \frac{M_r}{n} \times 100\% \tag{3}$$

$$\text{Minimize } RMSE = \sqrt{\frac{\sum_{i=1}^{n}(q_i - r_i)^2}{n}} \tag{4}$$

where, M_{pd}, is the number of events satisfying acceptable peak discharge; M_{pt} is the number of events satisfying acceptable peak runoff time; M_r is the number of events satisfying acceptable total runoff; n is the number of measurements data (input data); q_i is a obtained observation data by Korea Meteorological Administration (KMA) and ri is simulated data.

Table 1. The combination objective functions of 6 scenarios in this study

Scenario cases	Combination objective functions
1	Equations (1) and (2)
2	Equations (1) and (3)
3	Equations (1) and (4)
4	Equations (2) and (3)
5	Equations (2) and (4)
6	Equations (3) and (4)

2.3   Multi-objective Harmony Search Algorithm

The Harmony Search Algorithm (HSA) introduced by Geem et al. (2001) and Kim et al. (2001) proposed an MHOA based on the musical performance processes that occur when a musician searched for a better state of harmony, such as during jazz improvisation. HSA uses several parameters to find the optimal solution, such as Harmony Memory (HM), Harmony Memory Considering Rate (HMCR), Pitch Adjusting Rate (PAR) and

Bandwidth (BW). HSA have been used three main operators used here, Random Selection (RS), Memory Consideration (MC), and Pitch Adjustment (PA)—to seek better solutions from the previous HM. Based on these features, HSA can find a suitable sound or harmony (a local optimum) and eventually reach an aesthetically pleasing sound or harmony (a global optimum). Recently, multi-objective versions of HS have been proposed that employ non-dominated sorting and crowding-distance considerations within the standard HS framework.

A multi-objective harmony search (MOHS) has several features suited to these various applications, including a fast non-dominated sorting method, improved diversity and convergence (because of an implicit elitist selection method based on the Pareto dominance rank and a secondary selection method based on the crowding distance), and a constraint-handling technique for addressing constrained problems efficiently and supporting real coding representations. Firstly, standard MOHS, which uses non-dominated sorting and the crowding distance concept (Deb et al. 2002), was proposed with the standard HSA in this study.

3 Application Results

3.1 Study Area

The proposed model was applied to the Daecheong dam watershed in South Korea with its drainage area of 4,134 km^2. Daily precipitation, and evaporation data were collected by Korea Meteorological Administration (KMA). The observation data (Fig. 1) for 2 years (from 1987 to 1988) were used in this study.

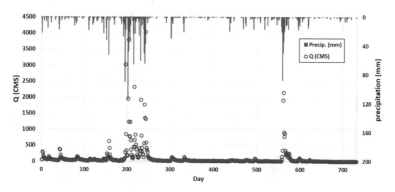

Fig. 1. Measurements data in the Daecheong dam watershed (Red: daily hyetograph; and Blue: daily hydrograph)

3.2   Results of Modified Tank Model

Figure 2 shows the best solution's Pareto front solution about peak discharge and RMSE among five optimization runs. In this study, the MOHS is used to the parameters calibration in modified tank model. Applied parameters in HS on the proposed model are HM size is 50, HMCR = 0.8, PAR = 0.3 and number of iterations = 50,000.

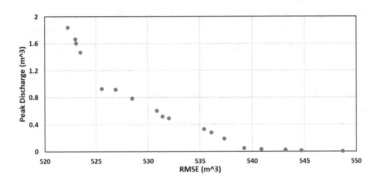

Fig. 2. Pareto fronts of best solution about peak discharge and RMSE (Scenario 1)

In conclusion, as an example to demonstrate that the calibrated modified tank model used multi-objective function accurately generates hydrographs close to the observed hydrographs, the calibrated results for the year 1987 to 1988 using the modified tank model are shown in the Fig. 3. The simulated results and observed runoff data for the proposed tank model are compared in this study. It can successfully simulate the runoff with little error and general watershed characteristics can be represented by calibrated parameters for longer periods.

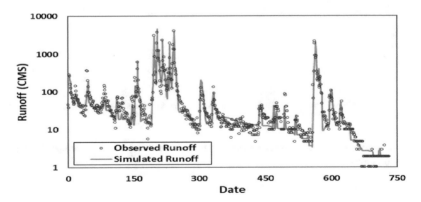

Fig. 3. The comparison between simulated runoff and measurement data for proposed tank model (y axis is in a logarithm scale) (scenario 1)

4 Conclusions

This study introduces various balancing approaches for the exploration and exploitation in the multi-objective automatic calibration for hydrological model. The proposed tank model with six different scenarios (combined scenarios by four objective functions) in a hydrological characteristic is estimated to compare the measurements data and simulation results. The proposed tank model has composed parameter calibration of four storage types of tank model with 14 parameters. In addition, we are set the combination of four objective functions (i.e., peak discharge; peak runoff time; total runoff; RMSE) as six different scenarios. In scenario 1, the error deviation of calibrated results shows relatively small in general data sets, whereas its shows much bigger in extreme data sets. The modified tank model under six different scenarios could be a successful alternative rainfall-runoff simulation model with its improved accuracy and Pareto solutions.

Acknowledgment. This work was supported by a grant from The National Research Foundation (NRF) of Korea, funded by the Korean government (MSIP) (No. 2016R1A2A1A05005306).

References

Boyle, D.P., Gupta, H.V., Sorooshian, S.: Toward improved calibration of hydrological models: combining the strengths of manual and automatic methods. Water Resour. Res. **36**(12), 3663–3674 (2000)

Cheng, C.T., Ou, C.P., Chau, K.W.: Combining a fuzzy optimal model with a genetic algorithm to solve multiobjective rainfall–runoff model calibration. J. Hydrol. **268**, 72–86 (2002)

Cheong, C.T., Wu, X.Y., Chau, K.W.: Multiple criteria rainfall-runoff model calibration using a parallel genetic algorithm in a cluster of computers. Hydrol. Sci. J. **50**(6), 1069–1087 (2005)

Deb, K., Pratap, A., Agarwal, S., Meyarivan, T.: A fast and elitist multi-objective genetic algorithm: NSGA-II. IEEE Trans. Evol. Comput. **6**(2), 182–197 (2002)

Geem, Z.W., Kim, J.H., Loganathan, G.V.: A new heuristic optimization algorithm: harmony search. Simulation **76**(2), 60–68 (2001)

Gupta, H.V., Sorooshian, S., Yapo, P.O.: Toward improved calibration of hydrologic models: multiple and noncommensurable measures of information. Water Resour. Res. **34**(4), 751–763 (1998)

Holland, J.H.: Adaptation in Natural and Artificial Systems. The University of Michigan Press, Ann Arbor (1993)

Jakeman, A.J., Hornberger, G.M.: How much complexity is warranted in a rainfall-runoff model? Water Resour. Res. **29**, 2637–2649 (1993)

Kim, J.H., Geem, Z.W., Kim, E.S.: Parameter estimation of the nonlinear muskingum model using harmony search. J. Am. Water Resour. Assoc. **37**(5), 1131–1138 (2001)

Yapo, P.O., Gupta, H.V., Sorooshian, S.: Multi-objective global optimization for hydrologic models. J. Hydrol. **204**, 83–97 (1998)

Application of Artificial Neural Network for Cyber-Attack Detection in Water Distribution Systems as Cyber Physical Systems

Kyoung Won Min[1], Young Hwan Choi[2],
Abobakr Khalil Al-Shamiri[2], and Joong Hoon Kim[3(✉)]

[1] Department of Civil, Environmental and Architectural Engineering,
Korea University, Seoul, South Korea
[2] Disaster and Science Technology, Korea University, Seoul, South Korea
[3] School of Civil, Environmental and Architectural Engineering,
Korea University, Seoul, South Korea
jaykim@korea.ac.kr

Abstract. Cyber-physical systems (CPSs) have been applied to water distribution systems for the efficient system operation and maintenance as an application field of fourth industrial revolution. Although CPSs technology is an efficient technology based on information communication technology, it is vulnerable to cyber-attack due to information disturbance of sensor and communication server. These cyber-attack causes serious problems in the operation of the water distribution systems such as decreases of water supply, water pollution and damage to physical systems. Therefore, a few studies were performed to organize the various cyber-attack scenarios and cyber-attack detection algorithms were developed. In this study, an artificial neural network model is applied for the detection of cyber-attack by varying the number of neurons. The developed detection technique can be contributed to the establishment of reliable the water distribution systems in real operation.

Keywords: Artificial neural network · Cyber-attack detection · Optimization · Neurons

1 Introduction

Cyber-physical systems (CPSs) can be defined as a combination of cyber network and physical device. It can monitor and control the physical processes in real time. CPSs have been applied to existing water infrastructures which two important components are the programmable logic controllers (PLCs) and supervisory control and data acquisition (SCADA) system. PLCs which communicate for actuators control are connected between sensors and actuators and. SCADA is a computer which supervise the operations of the entire water infrastructures and to store and analyze real-time data. Although CPSs technology is an efficient technology based on information communication technology, it is vulnerable to cyber-attack due to information disturbance of

© Springer Nature Switzerland AG 2020
J. H. Kim et al. (Eds.): ICHSA 2019, AISC 1063, pp. 82–88, 2020.
https://doi.org/10.1007/978-3-030-31967-0_10

sensor and communication server. Cyber-attack is an attempt to intentionally alter, disturb, or destroy information or programs transmitted or stored through a computer system or the network itself or through such networks. Generally, cyber-attack damages to the physical and communication devices (i.e., PLC, SCADA, sensor, and actuator) to provide wrong information related to the operation and management. Recently, WDSs also have been applied CPSs for efficient operation and management, and it also exposed the cyber-attack. If WDSs have the cyber-attack, it makes the degradation of water quality and the disturbance of water supply, and it can be spread great impact on local communities and economy. Because the water distribution system is an important infrastructure of the country, it is necessary to detect and take action against this cyber-attack in order to maintain normal operation of water distribution systems.

Therefore, this study proposes the cyber-attack detection technique using an artificial neural network. To detect the attack situation, the normal state data which does not contain any cyber-attack and abnormal state data which contain several attacks train the neural network, and the evaluation of detection performs considering the true and false alarms. Moreover, to improve detection performance, the parameters of the artificial neural network have been performed the sensitivity analysis for the number of neurons, we designed the optimal artificial neural network. The developed detection technique can be contributed to the establishment of the reliable water distribution systems in real operation.

2 Methodology

2.1 Cyber-Attack Specifications

WDSs as CPSs consist of a combination of physical systems such as pumps, valves, tanks and cyber systems such as PLCs, SCADA. When WDSs as CPSs are attacked, as shown in Fig. 1, the cyber-attack layer can be divided into five. Layer 1 means that actuators such as pumps and valves and sensors are directly attacked. If actuators such as pumps and valves malfunction, there is a problem that the water level in the tank goes up or down, the water supply on demand nodes does not work properly, and so on. Layer 2 is a connection link attack between PLC and actuators such as pumps and valves and sensors. Attacking causes malfunctions of actuators and sensors. Layer 3 means a connection link attack between PLCs. By this attack, each PLC malfunctions actuator or sensor due to erroneous information. Layer 4 means a connection link attack between SCADA and PLC. Layer 5 means attacking SCADA itself. This attack will result in the full unusual operation of WDS.

The scenario for these attacks can be divided into hydraulic and water quality criteria. In the hydraulic criteria, CPSs can malfunction through cyber-attack causing tank overflow or low level in a tank, lack of demand at some nodes. In water quality criteria, Pollutants may be injected by an attacker, which can deteriorate water quality. In addition, the age of water may also be deteriorated by the change of hydraulic elements such as pressure and flow rate by cyber-attack. This paper considers only the water age of water quality criteria and hydraulic criteria.

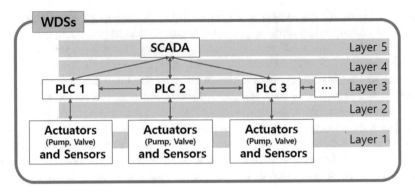

Fig. 1. Cyber-attack layer model for SCADA of WDSs as CPSs

2.2 Performance Indices

In this study, to evaluate the detection performance, true and false alarms are considered (Taormina et al. 2018). This formula accounts for both correct detection and false alarm. The success of detection (Sd) varies between 0 and 1. If the value of S is closing to 1, it means the perfect classification.

$$\text{Sd} = \frac{TPR + TNR}{2} \tag{1}$$

where, TPR is True Positive Rate and calculated as following: $TPR = TP/(TP + FN)$, it is the ratio of the cyber-attack that was really cyber-attacked when the result came out. TP and FN are the number of True Positives and False Negatives, respectively. Also, TNR is True Negative Rate and calculated as following: $TNR = TN/(TN + FP)$, it is the ratio of not cyber-attack that was not really cyber-attacked when the result came out. TN and FP are the number of True Negatives and False Positives.

2.3 Artificial Neural Network

Artificial Neural Networks (ANNs) is useful tools that have been applied in a variety range of field. ANNs is computing systems inspired by the biological neural networks that constitute animal brain. An artificial neural network is a model that has the ability to solve problems by changing the synaptic bond weight through artificial neurons that form a network by synaptic connections. These artificial neural networks are useful for solving nonlinear problems and have advantages such as less time and cost than statistical methods. The performance of artificial neural network depends on the weights of network, the number of neurons and the number of hidden layers.

3   Model Formulation

3.1   Study Network

C-Town WDS which is based on a real-world medium-sized network was used in this study (Taormina et al. 2017). This network has 388 demand nodes, 429 pipes, 7 tanks, 11 pumps and 4 valves. As depicted in Fig. 1, pumps, valves, and tank water level sensors are connected to nine PLCs, which are located in the proximity of the hydraulic components they monitor and control. There is also a single SCADA system that collects the readings from all PLCs and coordinates the operations of the entire network (Fig. 2).

Fig. 2. C-Town network

3.1.1   Datasets of Cyber-Attacks on C-Town

The applied dataset consists of the two training data, the test data. These datasets are generated by EPANET. The datasets are composed the water level in the tanks (7 variables), flow passing through pumps and actionable valves (12 variables), a status of pumps and actionable valves (12 variables) and pressure at inlet and outlet of pumping stations or actionable valves (12 variables). In order to obtain the pressure and flow information of the pump and the valve, the node ID is 14, 206, 256, 269, 280, 289, 300, 302, 307, 317, 415, 422 nodes were used. A total of 7 tank information was used for tank level information and a total of 12 pumps and valve information used for pump and valve status information.

Moreover, ANNs are trained by Training dataset 1 which does not contain any attacks and was generated from a one-year long simulation and Training Dataset 2 which contains several attacks and was generated from a 6 months long simulation. Then ANNs was tested by the Test dataset which contains several attacks but no labels and was generated form 3 months. This process progresses according to the number of neurons and the number of hidden layers, and then an optimal artificial neural network that is more accurate and faster to calculate can be constructed (Fig. 3).

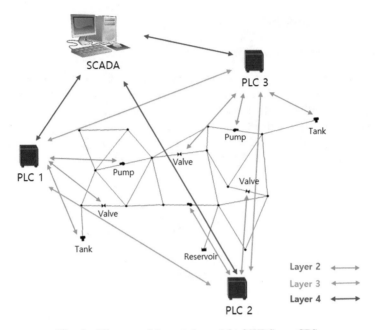

Fig. 3. Diagram of the attack model of WDSs as CPSs

4   Result

This study uses the success of detection to evaluate detection performance applying for the various the number of neurons which is one of the parameters in ANN. The number of neurons was changed from one to twenty and the detection performance represented in Table 1. Table 1 shows the value of Sd and the other results from ANN. The value of Sd has a range between 0 and 1, the closer to 1 means a perfect classification. According to the sensitivity analysis of the neurons number, when the number of neurons is 4, the detection performance shows the best result. It means the number of neurons is 4, the ability to detect cyber-attack is maximized. The Sd is calculated by TPR and TNR. In the case of 4 number of neurons, TNR shows a similar value as 0.8 but the TPR represents the highest value. Moreover, the probability of detection is about 80% regardless of the true positive rate and the true negative rate. From this result, the true positive rate occupies a larger proportion to improve detection performance.

Figure 4 shows the comparison results between the success of detection and the number of neurons in the whole simulation times. In this figure, the blue lines represent the actual attack scenarios according to time steps and the orange lines represent the success of detection. As the previous results mention, the best detection performance shows when the number of neurons is 4, as calculated from Table 1. Comparison of the other results, the detection probability of the true positive rate is better, but there are still many possibilities to improve the detection performance.

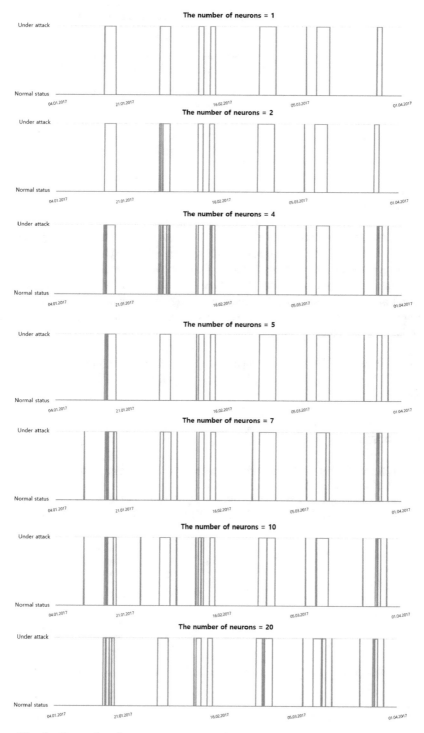

Fig. 4. Comparison between the success of detection and the number of neurons

Table 1. Artificial neural network performance depending on the number of neurons.

The number of neurons	Sd	TPR	TNR	TP	FP	TN	FN
1	0.400	0	0.08	0	407	1,680	2
2	0.736	0.60	0.80	4	403	1,680	2
4	**0.814**	**0.81**	**0.81**	**22**	**385**	**1,677**	**5**
5	0.670	0.54	0.80	5	401	1,677	5
7	0.732	0.65	0.81	17	390	1,673	9
10	0.690	0.57	0.80	15	392	1,671	11
20	0.780	0.75	0.81	18	389	1,676	6

5 Conclusion

Previous studies have been focused on smart city design only. However, in preparation for the 4th industrial revolution era, the security problems of water distribution systems such as cyber-attack should be prepared because WDSs is the important national infrastructure. By studying about security of WDSs, it can contribute to the establishment of safe water supply infrastructure throughout the whole process of designing and operating WDSs as CPSs. And then, cyber-attack scenarios need to be analyzed more broadly and more precise.

Acknowledgements. This work was supported by a grant from The National Research Foundation (NRF) of Korea, funded by the Korean government (MSIP) (No. 2016R1A2A1A 05005306).

References

Taormina, R., et al.: Characterizing cyber-physical attacks on water distribution systems. J. Water Resour. Plan. Manage. **143**(5), 04017009 (2017)

Taormina, R., et al.: Battle of the attack detection algorithms: disclosing cyber attacks on water distribution networks. J. Water Resour. Plan. Manage. **144**(8), 04018048 (2018)

Multi-objective Jaya Algorithm for Solving Constrained Multi-objective Optimization Problems

Y. Ramu Naidu[1], A. K. Ojha[1(✉)], and V. Susheela Devi[2]

[1] School of Basic Sciences, Indian Institute of Technology Bhubaneswar,
Bhubaneswar, Odisha, India
y.ramunaidu@gmail.com, akojha@iitbbs.ac.in
[2] Department of Computer Science and Automation, IISc, Bangalore, India
susheela@iisc.ac.in

Abstract. Solving multi-objective optimization problems (MOPs) is a challenging task since they conflict with each other. In addition, incorporation of constraints to the MOPs, called CMOPs for a short, increases their complexity. Traditional multi-objective evolutionary algorithms (MOEAs) treat multiple objectives as a whole while solving them. By doing so, fitness assignment to each individual is difficult. In order to overcome this difficulty, in this paper, multiple populations are considered for multiple objectives to optimize simultaneously and a hybrid method of Jaya algorithm (JA) and quasi reflected opposition based learning (QROBL), to maintain diversity among populations, is used as an optimizer to solve CMOPs. An archive is also used to store all non-dominated solutions and to guide the search towards the Pareto front. A local search (LS) method is performed on the archive members to improve their quality and converge to the Pareto front. The whole above process is named as CMOJA. The obtained results are compared with the state-of-the-art algorithms and demonstrated that the proposed hybrid method has shown its superiority to its competitors.

Keywords: Jaya algorithm ·
Quasi reflected oppositional based learning ·
Constrained multi-objective optimization problems

1 Introduction

A general mathematical form of CMOPs is given below:

$$f(x) = [f_1(x), f_2(x), \ldots, f_m(x)]$$
$$\text{subject to } g_i(x) \leq 0, \; i = 1, 2, \ldots, p \tag{1}$$
$$h_j(x) = 0, \; j = 1, 2, \ldots, q,$$

© Springer Nature Switzerland AG 2020
J. H. Kim et al. (Eds.): ICHSA 2019, AISC 1063, pp. 89–98, 2020.
https://doi.org/10.1007/978-3-030-31967-0_11

where $x = (x_1, x_2, \ldots, x_d)$ and $x_k \in [x_k^l, x_k^u]$, $k = 1, 2 \ldots, d$, x_k^l and x_k^u are lower and upper bounds of the decision variable x_k, respectively. Equation (1) has m objective functions, p inequality constraints and q equality constraints.

Solving CMOPs has been attracting researchers' attention for the past two decades and it has a great number of applications in many diverse fields such as Engineering design, Economics, Physics and Industry. Many evolutionary algorithms (EAs) have been implemented to deal with such design problems. The advantage of multi-objective approaches is that the decision maker can have a set of non-dominated solutions. Therefore he can choose a more suitable one based on his preference.

The main disadvantage of MOEAs is a selection of solutions for the next generation. We can not select them based on fitness like conventional EAs for solving single objective problems since all objectives conflict with each other. The search is inefficient unless the fitness is assigned to each solution properly. To address this issue, Zhan et al. [10] proposed a novel co-evolutionary method in which multiple populations are considered for multiple objectives (MPMO) to optimize simultaneously that means each population optimizes the corresponding objective function in each iteration by sharing search information among them. The particle swarm optimization (PSO) with modified velocity is adopted as an optimizer for each population. The velocity of each particle is modified for sharing search information from the archive which is used for storing all non-dominated solutions. Wang et al. [9] proposed co-operative differential evolution for solving multi-objective problems in which the same strategy, MPMO, is used for optimizing objectives. But an adaptive differential evolution is performed on each population. Naidu and Ojha [4] developed a hybrid invasive weed optimization for solving MOPs as well as systems of non-linear equations. It is named as HCMOIWO. In the HCMOIWO, a hybrid method of invasive weed optimization (IWO) and space transformation search performed on each population during the iteration process.

In this paper, a novel co-operative multi-objective jaya algorithm (CMOJA) is proposed to solve CMOPs. In the CMOJA, to address the fitness assignment problem and stagnation at local fronts, MPMO strategy is utilized to handle all objectives. The JA and QROBL are carried out for each population. The simplest static penalty method is used to handle the given constraints. An archive is also kept to accumulate all non-dominated solutions from all populations and it shares search information among populations to maintain diversity and approximate the whole Pareto-front. The LS [1] is performed on the archive members to improve their quality. At the end of the iteration, the farthest-candidate method (FCM) [1] is employed to limit the archive members if the archive has more members than its fixed limit.

The structure of this paper is as follows: In Sect. 2, the conventional JA and some basic definitions of opposition based learning method are reviewed. Section 3 presents the framework of the proposed method. In Sect. 4, some numerical examples and results are examined, and the obtained results are compared with well-known MOEAs. Section 5 summarizes the conclusions and some of our future research work.

2   Background

This section is devoted to furnish the information about the conventional JA and QROBL to understand the proposed method.

2.1   Jaya Algorithm

The recent population based meta-heuristic algorithm, Jaya Algorithm (JA), was proposed by Rao [5] which is inspired by the victory in a game or battle. Initially, it is used for solving unconstrained and constrained optimization problems later on many applications have been solved by using it due to its flexibility [3,6]. The advantages of the JA are that is parameter-free algorithm except algorithmic parameters such as population size and a termination condition. It has only one updating equation and defined as:

$$X_{i,j}(it+1) = X_{i,j}(it) + \overbrace{rand(0,1)(X_{b,j} - |X_{i,j}(it)|)}^{I} - \underbrace{rand(0,1)(X_{w,j} - |X_{i,j}(it)|)}_{II}$$

(2)

here it represents the current iteration and $rand(0,1)$ is an uniform random number lies between 0 and 1. $X_{b,j}$ and $X_{w,j}$ are the jth component of the best and worst individuals in the current population, respectively. In Eq. (2), the aim of the part I is the search towards the best individual and the part II is for avoiding the search towards the worst individual. If the new individual $X_i(it+1)$, is better than the old one $X_i(it)$ then the old one will be replaced by the new one. In order to understand the JA clearly, its flow chat is presented in Fig. 1.

2.2   Oppositional Based Learning (OBL)

The OBL, at first, was proposed by Tizhoosh [8] to increase the likelihood of finding better solutions. The definition of OBL is as follows: Let x be a decision variable in $[a, b]$ and x^o be its opposite, then

$$x^o = a + b - x.$$

(3)

Equation (3) can be extended to a vector in the d–dimensional space as

$$x_i^o = a_i + b_i - x_i, \quad x = (x_1, x_2, \ldots, x_d), \quad x_i \in [a_i, b_i], \quad i = 1, 2, \ldots, d.$$

(4)

Quasi-reflected opposition based learning (QROBL) is one of the variations of the OBL, proposed by Ergezer et al. [2] and it is defined as below:

Let $x = (x_1, x_2, \ldots, x_d)$ be real vector in the d–dimensional space and x^{qrf} be its quasi-reflected opposite on the same space. Thus

$$x_i^{qrf} = rand(x_i, c_i), \quad c_i = (a_i + b_i)/2, \quad i = 1, 2, \ldots, d.$$

(5)

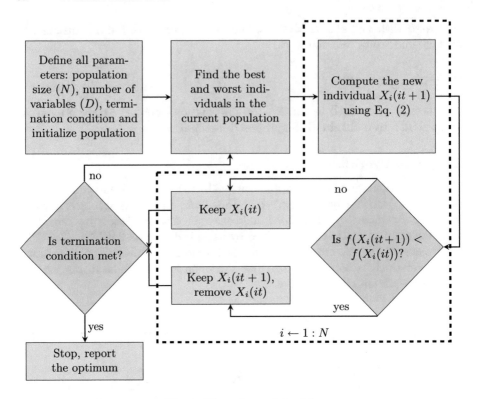

Fig. 1. Flow chart of the JA

3   Working Manner of the Proposed Hybrid Method (CMOJA)

In the MOEAs, a crucial operation is that maintaining an adequate trade-off between diversity and convergence of the algorithm. For this purpose, a hybrid method which combines the advantages of JA and QROBL is proposed to solve CMOPs. Furthermore, to avoid fitness assignment problem, multiple populations are considered for multiple objectives to be optimized simultaneously. In the CMOJA, an external archive is utilized to store all non-dominated solutions collected from all populations and all populations share their searching information through this archive. In order to improve the quality of solutions in the archive, the LS is employed on it.

The evolutionary process of the CMOJA is as follows: for a quick understanding, a schematic view of it is shown in Fig. 2. At first, m populations are initialized with uniform random numbers in the variable range where m is the number of objective functions. We consider the mth population to explain the evolutionary process of the CMOJA. QROBL is performed to generate opposition to each individual in the mth population. Then choose the best N individuals. An archive A is updated with non-dominated individuals extracted from all populations.

Here, the iterative process is started in which the best and worst individuals are found. Then the new individuals are computed by the following mathematical expression:

$$X_{i,j}(it+1) = X_{i,j}(it) + rand(0,1)(X_{b,j} - |X_{i,j}(it)|) - rand(0,1)(X_{w,j} - |X_{i,j}(it)|)$$
$$+ \underbrace{rand(0,1)(A_{k,j} - |X_{i,j}(it)|)}_{III}.$$

(6)

The part III encourages the search towards the Pareto front by using the sharing information of all populations from the archive A, and in which $A_{k,j}$ is the jth component of the archive member, selected randomly. The random selection of the archive member maintains the diversity with low computational cost. Then update the population. Next, the QROBL is performed on the new population and again update the population if the new population generated by QROBL is better. This process repeats for all populations according to their objective functions. Next, the LS is performed on the archive members. Now, accumulate all new populations, QROBL populations and new archive members. The next step is that update the archive A by extracting non-dominated individuals from the accumulation. Eventually, the FCM is activated if the size of A exceeds its considered limit. The reader can get information about the LS and FCM in [1]. The whole above process repeats until the termination condition is satisfied and the archive A is the desirable solution set. The Pseudo code of CMOJA is shown in Algorithm 1 for better understanding.

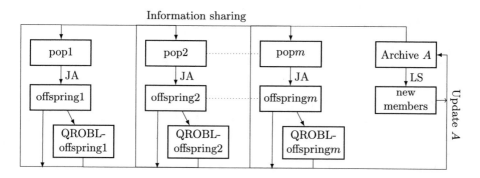

Fig. 2. A schematic view of CMOJA

4 Numerical Examples and Analysis of Results

To inspect the performance of the proposed method, three well-known CMOPs with diverse characteristics, consist of different Pareto fronts, are used. To compare the obtained results with that of state-of-the-art algorithms qualitatively

---

Algorithm 1.

Require: Define all fixed algorithmic parameters
  for $k = 1 \leftarrow m$ **do**
    for $i = 1 \leftarrow N$ **do**
      for $j = 1 \leftarrow D$ **do**
        $X_{i,j}^k = l_j + rand(0,1)(u_j - l_j)$
      end for
      Compute QROBL $(QROBL - offspring_{i,j}^k)$ corresponding to $X_{i,j}^k$
    end for
    Choose the best N individuals from X_i^ks and $QROBL - offspring_i^k$s for pop^k
  end for
  while $it < Maxit$ **do**
    for $k = 1 \leftarrow m$ **do**
      for $i = 1 \leftarrow N$ **do**
        Generate a new individual $(X_{new,i}^k)$ using Eq. (6) and evaluate it
        if $X_{new,i}^k$ is better than X_i^k **then**
          $X_i^k \leftarrow X_{new,i}^k$
        end if
        Compute a new QROBL individual $(QROBL - offspring_i^k)$ and evaluate it
        if $QROBL - offspring_i^k$ is better than X_i^k **then**
          $X_i^k \leftarrow QROBL - offspring_i^k$
        end if
      end for
    end for
    LS performs on the archive members
    Update the archive A
    if Size of A is greater than A_{limit} **then**
      FCM is activated to remove the extra members from A
    end if
  end while

---

and quantitatively, NSGA-II, MOPSO, and MWCA are employed. For this purpose, some of performance metrics (PMs): Generational Distance (GD), Metric of spacing (S), Metric of spread (Δ) are considered. For definitions and mathematical expressions of those metrics, readers can refer to [7].

The parameter set-up for the CMOJA is as follows: population size(N) is 10 and the size of the archive is 100. The CMOJA was programmed in MATLAB 2013Ra programming language. The results of NSGA-II, MOPSO and MWCA are taken directly from [7].

4.1 Constrained Multi-objective Problem 1: CONSTR

The CONSTR problem is taken from the literature [7]. It has two design variables and two inequality constraints. Its mathematical expression is as follows:

$$\text{Minimize } F(x) = \begin{cases} f_1(x) = x_1 \\ f_2(x) = \frac{1+x_2}{x_1} \end{cases} \tag{7}$$

$$\text{subject to } g_1(x) = 6 - (x_2 + 9x_1) \leq 0$$
$$g_2(x) = 1 + x_2 - 9x_1 \leq 0$$
$$0.1 \leq x_1 \leq 1, \ 0 \leq x_2 \leq 5.$$

For the CONSTR problem, the CMOJA uses 25,000 function evaluations for a fair comparison with its competitors. The statistical measurements, Mean and Standard Deviation (SD), of performance metrics are reported in Table 1. The obtained optimal Pareto front is depicted in Fig. 3a. From these results, it is inferred that the CMOJA outperformed its competitors in case of GD and S metric values in terms of Mean and SD, and in case of Δ, slightly worse than MWCA and better than NSGA-II and MOPSO. Less values of Mean of all PMs demonstrate the minimum distance from true Pareto front, good distribution and spread as well as the less values of SD demonstrate the robustness of the CMOJA.

Table 1. Comparisons of performance metrics for CONSTR problem.

Algorithm	GD		Δ		S	
	Mean	SD	Mean	SD	Mean	SD
NSGA-II	5.1349E−03	2.4753E−04	0.54863	2.7171EâĽŠ02	0.0437	0.0041
MOPSO	4.5437E−03	6.8558E−04	0.94312	3.6719E−01	N/A	N/A
MWCA	8.7102E−04	5.7486E−05	0.17452	2.0441E−02	0.0350	0.0010
CMOJA	2.1800E−04	9.9906E−06	0.19209	8.7986E−03	8.6263E−03	9.4716e−05

Note: N/A stands for Not Available

4.2 Constrained Multi-objective Problem 2: KITA

The KITA problem was formulated by Kita et al. [7] and has been used widely to examine the performance of optimization algorithms. It has two design variables and three inequality constraints, and it is a maximize type problem. The mathematical expression of the KITA is as follows:

$$\text{Maximize } F(x) = \begin{cases} f_1(x) = -x_1^2 + x_2 \\ f_2(x) = \frac{1}{2}x_1 + x_2 + 1 \end{cases} \tag{8}$$

$$\text{subject to} \quad g_1(x) = \frac{1}{6}x_1 + x_2 - \frac{13}{2} \leq 0$$

$$g_2(x) = \frac{1}{2}x_1 + x_2 - \frac{15}{2} \leq 0$$

$$g_3(x) = 5x_1 + x_2 - 30 \leq 0,$$

$$0 \leq x_1, \ x_2 \leq 7.$$

For this problem, 5000 function evaluations are considered for all optimization algorithms. The Mean and SD values of 30 independent runs for all PMs are listed in Table 2. From Table 2 and Fig. 3b, it is concluded that the MQRJA outperformed its competitors in terms of Mean and SD of all PMs. The optimal Pareto front of the KITA generated by CMOJA is illustrated in Fig. 3b. From this figure, we can observed that CMOJA achieved more non-dominated points than the true Pareto front.

Table 2. Comparisons of performance metrics for KITA problem.

Algorithm	GD		Δ		S	
	Mean	SD	Mean	SD	Mean	SD
NSGA-II	0.04	0.044	0.7863	0.1951	0.1462	0.1518
MOPSO	0.0467	0.0535	0.9925	0.1176	0.3184	0.4778
MWCA	0.0049	0.0045	0.3764	0.0744	0.0485	0.0478
CMOJA	7.1648E−04	6.4756E−05	0.2101	0.0331	0.0044	0.0011

Note: N/A stands for Not Available

4.3   Constrained Multi-objective Problem 3: SRN

The SRN problem was proposed by Srinivas and Deb [7]. It has two decision variables and two inequality constraints and its mathematical expression is given below:

$$F(x) = \begin{cases} f_1(x) = 2 + (x_1 - 2)^2 + (x_2 - 1)^2 \\ f_2(x) = 9x_1 - (x_2 - 1)^2 \end{cases} \tag{9}$$

$$\text{subject to} \quad g_1(x) = x_1^2 + x_2^2 - 225 \leq 0$$

$$g_2(x) = x_1 - 3x_2 + 10 \leq 0,$$

$$-20 \leq x_1, \ x_2 \leq 20.$$

In order to solve the SRN problem, 25,000 function evaluations are used as a termination condition. The obtained results, Mean and SD values of PMs for 30 runs, are reported in Table 3. From this Table, it is inferred that CMOJA has better values for GD and S than its competitors, but in case of Δ, it is better than NSGA-II and MOPSO whereas worse than MOWCA. The optimal Pareto front attained by CMOJA is portrayed in Fig. 3c in which the obtained non-dominated solution are very close to the true Pareto front.

Table 3. Comparisons of performance metrics for SRN problem.

Algorithm	GD		Δ		S	
	Mean	SD	Mean	SD	Mean	SD
NSGA-II	3.7069E−03	5.1034E−04	0.3869	2.5115E−02	1.5860	0.1337
MOPSO	2.7623E−03	2.0794E−04	0.6655	7.2196E−02	N/A	N/A
MWCA	2.5836E−02	5.0102E−03	0.1477	1.3432E−02	0.4164	0.0779
CMOJA	9.8942E−05	2.0101E−05	0.2081	0.0165	0.2416	0.0102

Note: N/A stands for Not Available

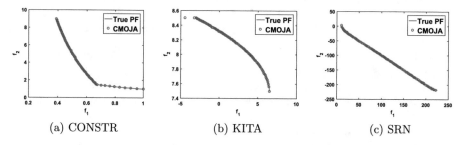

(a) CONSTR   (b) KITA   (c) SRN

Fig. 3. The optimal Pareto front obtained by CMOJA

5   Conclusions

This paper has developed a new version of multi-objective JA, named as CMOJA, for solving CMOPs. In the CMOJA, different populations optimize different objectives simultaneously by co-operating with each other. JA and QROBL are performed on each population to update it. The LS is also performed on the archive members to improve the quality and converge to the Pareto-front. The FCM is used to eliminate the extra members of the archive. In order to test the performance of the CMOJA, three CMOPs are solved and the obtained results are compared with that of well-known MOEAs, NSGA-II, MOPSO, and MOWCA. It is inferred that the CMOJA is superior to its competitors in most of problems.

In the future, the CMOJA can be studied deeply and developed for many objective problems and some multi-objective engineering design problems. Furthermore, it will also be used for solving systems of non-linear equations.

References

1. Chen, B., Zeng, W., Lin, Y., Zhang, D.: A new local search-based multiobjective optimization algorithm. IEEE Trans. Evol. Comput. **19**, 50–73 (2015)
2. Ergezer, M., Simon, D., Du, D.: Oppositional biogeography-based optimization. In: 2009 IEEE International Conference on Systems, Man and Cybernetics, pp. 1009–1011 (2009)

3. Huang, C., Wang, L., Yeung, R.S.C., Zhang, Z., Chung, H.S.H., Bensoussan, A.: A prediction model-guided Jaya algorithm for the PV system maximum power point tracking. IEEE Trans. Sustain. Energy **9**, 45–55 (2018)
4. Naidu, Y.R., Ojha, A.K.: Solving multiobjective optimization problems using hybrid cooperative invasive weed optimization with multiple populations. IEEE Trans. Syst. Man Cybern. Syst. **48**, 821–832 (2018)
5. Rao, R.: Jaya: a simple and new optimization algorithm for solving constrained and unconstrained optimization problems. Int. J. Industr. Eng. Comput. **7**, 19–34 (2016)
6. Rao, R.V., Rai, D.P., Balic, J.: A multi-objective algorithm for optimization of modern machining processes. Eng. Appl. Artif. Intell. **61**, 103–125 (2017)
7. Sadollah, A., Eskandar, H., Kim, J.H.: Water cycle algorithm for solving constrained multi-objective optimization problems. Appl. Soft Comput. **27**, 279–298 (2015)
8. Tizhoosh, H.R.: Opposition-based learning: a new scheme for machine intelligence. In International Conference on Computational Intelligence for Modelling, Control and Automation and International Conference on Intelligent Agents, Web Technologies and Internet Commerce (CIMCA-IAWTIC 2006), vol. 1, pp. 695–701 (2005)
9. Wang, J., Zhang, W., Zhang, J.: Cooperative differential evolution with multiple populations for multiobjective optimization. IEEE Trans. Cybern. **46**, 2848–2861 (2016)
10. Zhan, Z.-H., et al.: Multiple populations for multiple objectives: a coevolutionary technique for solving multiobjective optimization problems. IEEE Trans. Cybern. **43**, 445–463 (2013)

DSP-Based Implementation of a Real-Time Sound Field Visualization System Using SONAH Algorithm

Zhe Zhang[1,2], Ming Wu[1,2], and Jun Yang[1,2]([✉])

[1] Institute of Acoustics, Chinese Academy of Sciences,
No. 21 North 4th Ring Road, Haidian District, Beijing 100190, China
jyang@mail.ioa.ac.cn
[2] University of Chinese Academy of Sciences,
No. 19(A) Yuquan Road, Shijingshan District, Beijing 100049, China

Abstract. Many near-field acoustical holography (NAH) algorithms, including statistically optimized near-field acoustical holography (SONAH), have been well developed and verified via computer simulations. In this paper, design, implementation, and performance of a DSP-based real-time system for sound field visualization using SONAH algorithm is discussed. The prototype system consists of a 24-element microphone array connected to a TMS320C6678 DSP daughterboard hosted by a PC, validated by both computer simulation and experiments performed in an anechoic chamber, visualizing the sound field of a reconstruction plane in front of two speakers. The system has prospective practical applicability in industry applications.

Keywords: Digital signal processor (DSP) ·
Statistically optimized near-field acoustical holography (SONAH) ·
Microphone array

1 Introduction

Statistically optimized near-field acoustical holography algorithm manages to avoid the errors and limitations of other NAH algorithms caused by using spatial DFT [1], also reducing the side-effects caused by the measuring aperture size of microphone array. During past twenty years, various SONAH algorithms in different coordinates based on sensor array were attempted to achieve higher resolution and higher robustness in various environments [2–6]. However, very few describe the actual implementation of acoustical holography system [7, 8] and its use in practical real-time sound field visualization. There are also papers discussing the design, implementation, and performance of DSP system with microphone array for other purposes like DOA estimation and beamforming [9, 10]. The system described in this paper uses TMS320C6678 multi-core digital signal processor as a core processor, cooperating with other associated hardware circuits. Taking advantages of the KeyStone multi-core architecture of the processor, the hardware and software are developed to run different tasks on different cores, thus providing a real-time performance of manipulating data and calculation of SONAH algorithm on a wide frequency band with low latency. After the reconstructed sound field were computed, the

© Springer Nature Switzerland AG 2020
J. H. Kim et al. (Eds.): ICHSA 2019, AISC 1063, pp. 99–110, 2020.
https://doi.org/10.1007/978-3-030-31967-0_12

data were uploaded to the host PC in a one-frame delay, which equals 16 ms under a 16000 Hz sampling rate and 256 samples buffer. The host PC then visualizes the reconstruction data. The prototype system is tested via computer simulation and experiments conducted in the free-field environments of an anechoic chamber.

Sound imaging techniques implemented by the system are becoming increasingly popular in various field, such as NVH problems in vehicle design, measuring of source positions, behavior, intensity and physical insight with details, evaluation of the performance of a speaker system, and so on [11]. There are also prospective applications in art field. Considering the more attention on "audiovisual" from visual and sound artists, the concept of "acoustic-visual" may be developed.

2   SONAH Algorithm

In this section, a brief review on SONAH algorithm [12] is presented with a data model adopted in following implementation.

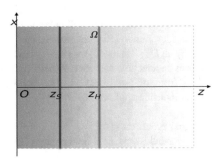

Fig. 1. Sound source and sound field in a SONAH model.

Consider a sound field reconstructing problem under free field condition. The surface of sound source is located at $z = z_S$ and region Ω is a source-free field occupied by air, shown as Fig. 1. According to the wave equation and spatial Fourier transform, sound pressure of a point $r = (x, y, z)$ can be represented as:

$$p(\mathbf{r}) = \frac{1}{4\pi^2} \iint P(k_x, k_y, z_s) e^{i[k_x + k_y + k_z(z-z_s)]} dk_x dk_y \tag{1}$$

which indicating that sound pressure at point r can be represented as the sum a series of elementary waves propagating from the sound source $z = z_S$. Every elementary wave has its corresponding wave number vector and weighting coefficient and can be also represented as:

$$\Phi(\mathbf{k}, \mathbf{r}) = F(k_z) e^{i[k_x + k_y + k_z(z-z_s)]} \tag{2}$$

where $\mathbf{k} = (k_x, k_y, k_z)$ is the wave-number vector and $F(k_z)$ is the amplitude weighting function. We can also see from Eq. (2) that $\Phi(\mathbf{k}, \mathbf{r})$ contains both the propagation waves inside the radiation circle and the evanescent waves outside the radiation circle from the source.

Considering the amplitude weighting function $F(k_z)$, Hald J. proposes $F(k_z) = F_0\sqrt{1/|k_z|}$ as the "omni-directional weighting" [13], providing a better accuracy for sound waves with a large incident angles. In this paper, we choose the unit weighting, $F(k_z) = 1$, for the sake of experiment conditions discussed in next section.

Implement a wave-number domain sampling with Eqs. (1) and (2), we yield:

$$p(\mathbf{r}) = \sum_n \frac{\Delta k_x \Delta k_y}{4\pi^2} P(\mathbf{k}_n, z_s)\Phi(\mathbf{k}_n, \mathbf{r}) \tag{3}$$

Then assume that the sound pressure on the holography plane $z = z_H$ has been measured at a set of positions. Apply Eq. (3):

$$p(\mathbf{r}_{Hj}) = \sum_n \frac{\Delta k_x \Delta k_y}{4\pi^2} P(\mathbf{k}_n, z_s)\Phi(\mathbf{k}_n, \mathbf{r}_{Hj}) \tag{4}$$

where

$$\mathbf{r}_{Hj} = (x_j, y_j, z_H) \; (j = 1, 2, \cdots, M)$$

Similarly, according to the elementary wave model, the elementary waves at $\mathbf{r}$ can be represented as a linear combination of corresponding elementary waves at the sampling points on the holography plane [13]:

$$\Phi(\mathbf{k}_n, \mathbf{r}) = \sum_{j=1}^{M} c_j(\mathbf{r})\Phi(\mathbf{k}_n, \mathbf{r}_{Hj}) \tag{5}$$

where $c_j(\mathbf{r})$ is the weighting function of the elementary waves.

Combining Eqs. (3), (4), and (5), we can yield:

$$p(\mathbf{r}) = \sum_{j=1}^{M} c_j(\mathbf{r})p(\mathbf{r}_{Hj}) = \mathbf{p}_H^T \mathbf{c}(\mathbf{r}) \tag{6}$$

which means that we can derive the sound pressure at $\mathbf{r}$ in Ω as long as we obtain the weighting function $c_j(\mathbf{r})$.

To calculate $c_j(\mathbf{r})$, rewrite Eq. (5) in the form of matrix:

$$\boldsymbol{\alpha}(\mathbf{r}) = \mathbf{A}\mathbf{c}(\mathbf{r}) \tag{7}$$

where

$$\alpha(\mathbf{r}) = \begin{bmatrix} \Phi(\mathbf{k}_1, \mathbf{r}) \\ \Phi(\mathbf{k}_2, \mathbf{r}) \\ \vdots \\ \Phi(\mathbf{k}_{N-1}, \mathbf{r}) \end{bmatrix}, A = \begin{bmatrix} \Phi(\mathbf{k}_1, \mathbf{r}_{H1}) & \Phi(\mathbf{k}_1, \mathbf{r}_{H2}) & \cdots & \Phi(\mathbf{k}_1, \mathbf{r}_{HM}) \\ \Phi(\mathbf{k}_2, \mathbf{r}_{H1}) & \Phi(\mathbf{k}_2, \mathbf{r}_{H2}) & \cdots & \Phi(\mathbf{k}_2, \mathbf{r}_{HM}) \\ \vdots & \vdots & \ddots & \vdots \\ \Phi(\mathbf{k}_N, \mathbf{r}_{H1}) & \Phi(\mathbf{k}_N, \mathbf{r}_{H2}) & \cdots & \Phi(\mathbf{k}_N, \mathbf{r}_{HM}) \end{bmatrix}$$

Considering the Nyquist sampling condition and trade-off of the calculation load, the total sampling points for solve the equation is chosen as $N = (2N_x - 1)(2N_y - 1)$. Then we can derive from Eq. (7):

$$c(\mathbf{r}) = \mathbf{A}^+ \alpha(\mathbf{r}) \tag{8}$$

Thus, we can reconstruct the sound field in the field Ω with Eqs. (6) and (8).

3 Implementation

3.1 Hardware Design

The block diagram of the developed prototype system is depicted in Fig. 2. The TMS320C6678 processor plays the center role in the real-time operations. The 24-channel ADC chip transforms analog signals captured by the microphone array with 24 elemental condenser microphones on the cross-points of 5×5 grid except for the bottom-right corner, shown in Fig. 3 (left). The distance between the nearest microphones in the array is 0.1125 m. The DDR3 memory in the system makes it possible to store large floating number arrays of the coefficients downloaded from the host PC. The completed circuits board connected with the ADC chip is shown in Fig. 3 (right).

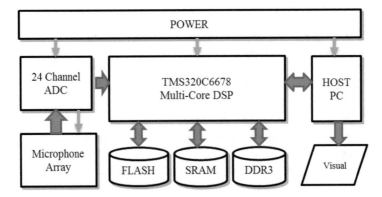

Fig. 2. Block diagram of the real-time sound field visualization system.

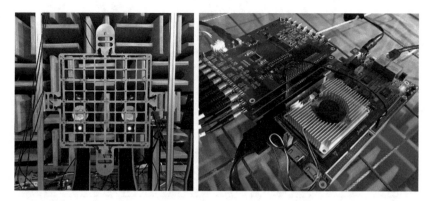

Fig. 3. The 24-channel microphone array (left) and the circuit board connected with the 24-channel ADC chip of the prototype system (right).

3.2 Software Development

Sound is captured by the microphone array and digitized by the ADC chip at the sampling rate of 16 kHz. Thanks to the highly performance of the TMS320C6678 processor and its KeyStone multi-core architecture, the master core of the processor plays the role of data retrieval and communicating with host PC and simultaneously the slave cores fulfill the tasks of reading the coefficient block in the memory, manipulating matrices, and communicating with master core. Once the system starts to run, a 256-sample frame of data from each channel is gathered and an FFT with Blackman window is performed on all 24 channels in the master core. The master core then sends the FFT data in each frequency (up to the 8 kHz in this 16 kHz-sampling case) to the corresponding slave core that takes charge of that frequency. While the slave cores are manipulating the calculations of algorithms in current frame, the master core reads and sends the calculated results of last frame that the slave cores have finished to the host PC. Then the loop starts over again when the timer calls the master core to work on the next frame.

In addition, the reconstruction coefficients calculated from Eq. (7) may be large in data size, especially when the frequency range is big and/or the reconstruction resolution is high. Thus, we take the advantages of the mass storage of DDR3 and the high speed of RAM. The data block will be moved into RAM whenever the certain coefficient array is in need.

By implementing such software scheme, we guarantee that the system works in real-time with a low latency level: one-frame delay (16 ms under the sampling rate and buffer size we use now). Anyway, there may also be latency caused by the analog-digital conversion and other process that should be taken into consideration in practical applications.

4   Performance

4.1   Computer Simulation

The SONAH algorithm described in Sect. 2 and prototype system described in Sect. 3 (except for the microphone array) were setup and run under the condition shown in Table 1.

Table 1. Computer simulation setup

z_S	z_H	Sound source #1 on z_H	Sound source #2 on z_H
$z_s = 0.1$ m	$z_H = 0.15$ m	(−0.1 m, −0.1 m)	(0.1 m, 0.1 m)

Two virtual sound sources were ideal point source located on the source plane $z = z_S$. The reconstruction plane was chosen as the same plane as the holography plane $z = z_H$ for the sake of convenience. The reconstruction resolution is set to 1 cm, which means there will be 43 * 43 pixels on the visualization image.

The simulation results were shown in Fig. 4, visualizing the "true" value of the sound pressure and reconstruction value of the sound pressure on the reconstruction plane, respectively. In Fig. 5 (right), we plot two curve surfaces in the same coordinate to directly compare the results of reconstruction with the "true", which indicates a consistent trend of sound pressure distributing on the reconstruction plane.

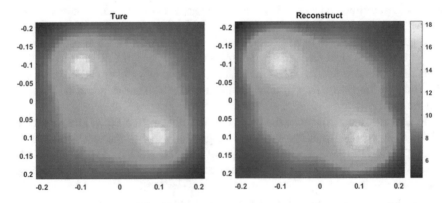

Fig. 4. Computer simulation results.

Figure 5 (left) shows the relative error of the reconstructing results compared with the "true" value. The maximum relative error is 30%, arising on the edge of the microphone array aperture. We find that when we choose a discrete representation in wave number domain in order to solve Eqs. (5) and (7), the solutions will lead to wrap-around errors just like in traditional NAH based on spatial DFT processing.

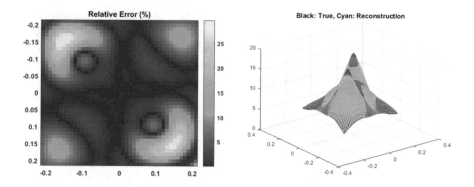

Fig. 5. Relative error of reconstruction (left) and Sound pressure curve surface of "true" value and reconstructed value (right).

4.2   Experiment Results

The system has been tested in an anechoic room at Institute of Acoustics, Chinese Academy of Sciences. The whole setup of the experiment is shown in Fig. 6 (left). We use a pair of YAMAHA HS5 speakers as the sound source which is located in front of the microphone array. According to the Nyquist theorem:

$$2\pi/\Delta \geq 2k_{\max} \tag{9}$$

Our microphone array with a sampling space of 0.1125 m is capable for the max frequency:

$$f_{\max} = c/2\Delta = 1528.88\,\text{Hz} \tag{10}$$

Typically, the distance between sound source and holography plane satisfies $d \leq 0.5\lambda_{min}$ [14]. Considering the exponential decay of the evanescent wave and the low noise level in this case, we go slight far from it and set the distance $d = 0.1$ m. The coordinate modeled based on the setup is shown in Fig. 6 (right). The speakers we use

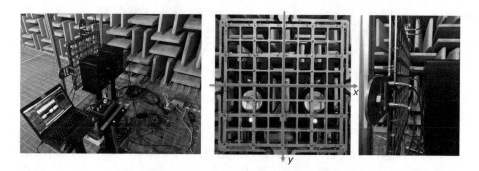

Fig. 6. Experiment setup and coordinates.

each has a woofer and tweeter taking charge of high frequency and low frequency, with a cross-point of 2 kHz. Thus, we expected four sound sources in total if we drive the speakers using a signal up to 2 kHz. The location of the centers of woofers and tweeters are depicted in Table 2. The reconstruction plane is coincided with the holography plane, same as the computer simulation.

Table 2. Experiment setup

Left woofer	Left tweeter	Right woofer	Right tweeter
(–0.17 m, 0.06 m)	(–0.17 m, –0.09 m)	(0.17 m, 0.06 m)	(0.17 m, –0.09 m)

Figure 7 shows the sound pressure distribution of 100 Hz sine wave on the reconstruction plane. The top-left image is the distribution when only the speaker on the left makes sounds while the top-right image only the speaker on the right. The bottom-left image shows the sound pressure distribution on the reconstruction plane when the two speakers play the same in-phase sine wave, while the bottom-right image shows the out-of-phase situation. Figures 8 and 9 also follow the same arrangement above, showing the distribution of 1000 Hz and 2000 Hz sine waves.

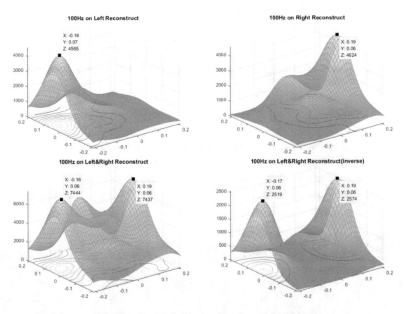

Fig. 7. Sound field visualization of 100 Hz.

In our visualized reconstruction image, we can clear see the distribution of sound pressure in front of the speakers. The peak amplitudes show the positions of center of the corresponding sound sources. When the speakers work on low frequency such as 100 Hz in Fig. 7 and 1000 Hz in Fig. 8, only the woofer vibrates and radiates sounds. In addition, comparing the in-phase sine waves sound field and the out-of-phase ones, the interference phenomenon is captured by our system: in the bottom-left images there are strengthened bands in the center while in the bottom-right images that bands are weakened.

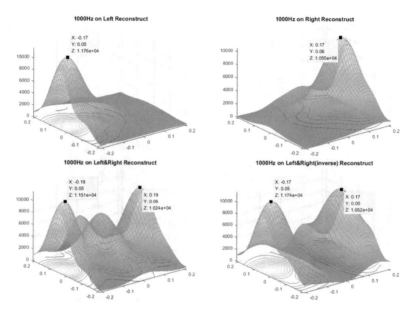

Fig. 8. Sound field visualization of 1000 Hz.

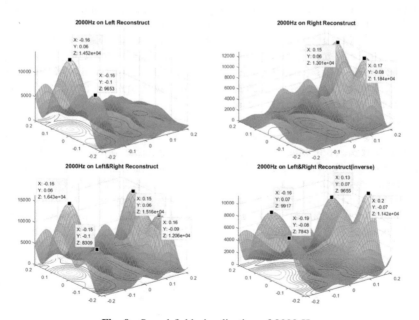

Fig. 9. Sound field visualization of 2000 Hz.

Although Eq. (10) tells that the prototype system has a cutoff frequency of about 1500 Hz, we still go slightly far beyond that to test the system up to 2000 Hz to see the expanding ability of it. Figure 9 shows the reconstruction sound pressure of 2000 Hz sine waves and Fig. 10 shows the reconstruction sound pressure in different layout of visualization of wide-band noise covering the frequency range of 60 Hz–2000 Hz, independent with each other from two speakers. We can see that the system still gets a good performance of orientation. And it meets our expectations that the speakers work on both woofer and tweeter when the frequency of the signal reaches the cross-point 2000 Hz.

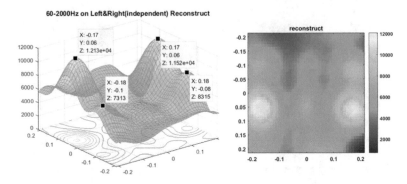

Fig. 10. Sound field visualization of independent wide-band noise (60 Hz–2000 Hz).

Finally, we evaluate the accuracy of sound source localization of the prototype system in Fig. 11, where the "X" marks represent the true location of the sound source in Table 2. And the "O" marks represent the localization results of our system (some results fall on the same coordinates so that the marks overlap with each other). The maximum error of localization is less than 5 cm and there may contain some measurement error due to the hands-on measuring and the vibration of the grid-floor of the anechoic chamber. The system performs better on low frequency than high frequency, which is also in line with our forecast.

Additionally, we also carry out some test focusing on the dynamic performance of the prototype system, including in-phase and out-of-phase sine waves sound field within a short period of time, wide-band noise with a linear amplitude-in-dB envelope, and stereo playback of a song clip of "In My Life" by The Beatles (chosen because the instruments and vocals were panned to hard left or hard right owing to the audio recording and mixing technology in the 60 s). The system visualizes the sound pressure of the reconstruction plane dynamically with an FPS of 62.5, offering a smooth visual effect. Because of the limitations of paper publication, we captured some results as GIF on a web which you can access in the link [15].

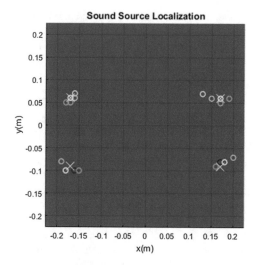

Fig. 11. Accuracy of sound source localization.

5   Conclusion

A DSP-based real-time sound field visualization system using SONAH algorithm is implemented. The prototype system is capable of visualizing the sound pressure distribution of the reconstruction plane in real-time. Computer simulation and free-field experiments have demonstrated that the prototype system is capable to visualize the sound field and locate the sound source within the cutoff frequency, also gains some extension ability to perform beyond the frequency range.

Further work will include refining the current system for better performance and extending the reconstruction plane from the holography plane. Other methods to solve Eq. (7) will also be implemented to remove the warp-around problem caused by the discrete representation in the wave-number domain. Furthermore, the software on the host PC will be developed and upgraded to offer more effective features for sound field visualization as well as valid user interaction.

References

1. Steiner, R., Hald, J.: Near-field acoustical holography without the errors and limitations caused by the use of spatial DFT. Int. J. Acoust. Vibr. **2**(6), 834–850 (2001)
2. Cho, Y., Bolton, J., Hald, J.: Source visualization by using statistically optimized near-field acoustical holography in cylindrical coordinates. J. Acoust. Soc. Am. **4**(118), 2355–2364 (2005)
3. Li, W.: Study on statistically optimal cylindrical near-field acoustical holography. Chin. J. Mech. Eng. **4**(41), 123 (2005)
4. Jacobsen, F., Jaud, V.: Statistically optimized near field acoustic holography using an array of pressure-velocity probes. J. Acoust. Soc. Am. **3**(121), 1550–1558 (2007)

5. Jean-Michel, F., Annie, R.: Time domain nearfield acoustical holography with three-dimensional linear deconvolution. J. Acoust. Soc. Am. **3**(143), 1672–1683 (2018)
6. Chaitanya, S.K., Thomas, S., Srinivasan, K.: Sound source localization using cylindrical nearfield acoustic holography. In: INTER-NOISE and NOISE-CON Congress and Conference Proceedings, InterNoise 18 (2018)
7. Hald, J.: Patch holography in cabin environments using a two-layer handheld array with an extended SONAH algorithm. In: Proceedings of Euronoise (2006)
8. Luo, Z.W., Comesana, D.F., Zheng, C.J., Bi, C.: Near-field acoustic holography with three-dimensional scanning measurements. J. Sound Vib. **439**, 43–55 (2019)
9. Rabinkin, D.V., Renomeron, R.J., Dahl, A.J., French, J.C., Flanagan, J.L., Bianchi, M.: A DSP implementation of source location using microphone arrays. J. Acoust. Soc. Am. **4**(99), 2503–2529 (1996)
10. Mahmod, A., Shatha, M.: TMS320C6713 DSP kit based hardware implementation for the microphone array beamforming system. Int. J. Comput. Appl. **6**(179), 40–45 (2017)
11. Scholte, R.: Fourier based high-resolution near-field sound imaging. Eindhoven Technische Universiteit (2008)
12. Hald, J.: Basic theory and properties of statistically optimized near-field acoustical holography. J. Acoust. Soc. Am. **125**, 2105–2120 (2009)
13. Chen, X., Bi, C.: Jin chang sheng quan xi ji shu ji qi ying yong. Ke xue chu ban she, Beijing (2013)
14. Bi, C., Chen, X., Zhou, R., Chen, J.: Reconstruction and separation in a semi-free field by using the distributed source boundary point method-based nearfield acoustic holography. J. Vib. Acoust. **3**(129), 323 (2007)
15. Sound Field Visualization. https://zhezhang.me/2019/03/sound-field-visualization-using-sonah-algorithm. Accessed 22 Mar 2019

The Economic Impact Analysis of the 1994 Northridge Earthquake by Deep Learning Tools

Zhengru Tao$^{(\boxtimes)}$ (iD), Lu Han, and Kai Bai

Key Laboratory of Earthquake Engineering and Engineering Vibration,
Institute of Engineering Mechanics, China Earthquake Administration,
Harbin 150080, China
taozr@foxmail.com

Abstract. Generally, a huge earthquake may be a big shock to the stock market, but it is difficult to measure it. We seek to quantify it in this paper. The normal price is predicted in the event window of the 1994 Northridge Earthquake, and then the abnormal return is obtained from the observed price. By testing the significant of abnormal returns, the shock can be quantified. Although numerous statistical and economic models have been developed to predict the movement of stock price, it is still a huge challenge. Rather than fitting the price to specific models, deep learning models of artificial neural network, namely a Long Short-Term Memory (LSTM) and a nonlinear autoregressive (NAR) neural network, are employed to predict the daily movements of US stock market in the event window. And, T-test and sign test are used to test the significance of the abnormal return in the event window. From the result of both two models, the impact of the Northridge Earthquake on the whole market is not significant at 95% confidence level; 20 stocks react positively, among which 5 stocks are from housing-related industry; 23 stocks react negatively, among which 8 stocks are from service industry.

Keywords: LSTM · NAR network · Northridge Earthquake · Stock market

1 Introduction

This is the 25th anniversary of the M6.7 Northridge Earthquake, which struck the Los Angeles region of southern California at 4:30 am on January 17, 1994. The epicenter was located in the San Fernando Valley, 1.5 km from the campus of California State University, Northridge, and about 32 km west-northwest of downtown Los Angeles [1], where is heavily populated and built-up, and around 10 million people were awakened by the shock. 57 people died, more than 9000 were injured, and more than 20000 were displaced from their homes by the effects of the quake [2]. The 10–20 s of strong shaking collapsed about 4,000 buildings, another 8,500 were moderately damaged, and more than 1600 were unsafe to enter. Seven major freeway bridges in the area collapsed, and 170 were damaged, disrupting traffic in the Ventura-Los Angeles region for weeks following the earthquake. Communication, water and power distribution systems were also affected, and ruptured gas lines exploded into fires [2, 3].

© Springer Nature Switzerland AG 2020
J. H. Kim et al. (Eds.): ICHSA 2019, AISC 1063, pp. 111–121, 2020.
https://doi.org/10.1007/978-3-030-31967-0_13

The intensity contour and the records on ShakeMap stations, given by the United States Geological Survey (USGS), are shown in Fig. 1 [4].

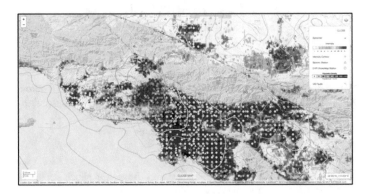

Fig. 1. The intensity contour, records on ShakeMap stations and population density, from USGS [4].

Heavy population was exposed to intense shaking in Fig. 1, where a large inventory of buildings and structures were concentrated. The residential damage estimated at more than $20 billion (nearly $35 billion in 2019 dollars) [5], the overall losses estimated at $44 billion, with insured losses estimated at $15.3 billion ($25.6 billion in 2017 dollars) [6]. It is the costliest earthquake in the U.S. [5].

Some researchers estimated economic losses from this quake during e following several years. After two months, it was not clear that how many businesses in the area of impact were damaged, or what kinds of damage and disruption were experienced [7]. In 1996, Tierney focused on the immediate and longer-term impacts the earthquake had on businesses in the Greater Los Angeles region, based on 1100 returned questionnaires [8]. The survey employed a three-stage sampling methodology, the sampling universe included all businesses in the cities of Los Angeles and Santa Monica. The survey obtained information on both direct physical damage to the business property and the disruption of utility services at the business site. Four years later, Eguchi et al. thought it was still difficult to answer the question of whether the accurate final dollar loss figure was completed [9]. Other researcher discussed the effect of this event on individual industries, like insurance, transportation and electricity [10–15].

Since the social-economic circumstance is more and more complicated, it is an impossible task to evaluate the impact of a destructive earthquake by tracing the direct loss, the secondary loss to the business interruption. On the other hand, the stock market can be a proxy, and the reaction of listed firms and investors to the earthquake can be observed in real time. So, the indices of US stock market and the price of individual stocks are monitored to measure the impact of the Northridge Earthquake in this paper.

2 Methodology, Data and Model Test

To measure the impact, the normal situation of the stock market should be settled, which means that the target earthquake does not occur. Thus, the "abnormal" return can be obtained and the significance can be tested. Two deep learning models of artificial neural network, LSTM and NAR network of recursive neural network (RNN), are employed to predict the normal situation in the following, which are popular for time series analysis, especially in the fields of economics and finance.

LSTM is one of recurrent network algorithms, which is a novel, efficient, gradient-based method, introduced by Hochreiter and Schmidhuber in 1997 [16]. It has been developed and applied continuously in recent years, especially for time series analysis [17–20]. It is a powerful time-series model, in which a memory cell c_j is structured as Fig. 2.

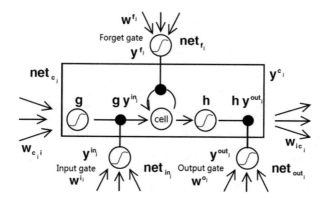

Fig. 2. Architecture of memory cell and three gate units, redrawn from [16].

In this module, there are five essential components. Input gate i_t decides the short-term memory stores in the cell state from the current input, which is calculated by Eq. (1). Forget gate f_t decides the long-term memory stores in the cell state from the previous cell state, which is calculated by Eq. (2). Output gate o_t decides the memory contents stores in the hidden state from the current cell state, which is calculated by Eq. (3).

$$i_t = \sigma\left(W_{xh}^i x_t + W_{hh}^i h_{t-1}\right) \tag{1}$$

$$f_t = \sigma\left(W_{xh}^f x_t + W_{hh}^f h_{t-1}\right) \tag{2}$$

$$o_t = \sigma\left(W_{xh}^o x_t + W_{hh}^o h_{t-1}\right) \tag{3}$$

where, W is the weight, which is a random value; x_t is the training data; h_{t-1} is the previous cell state.

A tanh layer creates a state g_t that could be added to the input gate, as Eq. (4).

$$g_t = \tanh\left(W_{xh}^g x_t + W_{hh}^g h_{t-1}\right) \tag{4}$$

By combining the forget gate and the input gate, an update to the state can be created as cell state c_t, which stores short-term and long-term memories.

$$c_t = c_{t-1} \cdot f_t + g_t \cdot i_t \tag{5}$$

Hidden state h_t decides the output state, which is the cell state passes through tanh and is multiplied by the output gate, as

$$h_t = \tanh(c_t) \cdot o_t \tag{6}$$

One of DNN, NAR network, is employed as the other option. The movement of the market is predicted by several last values, as Eq. (7).

$$y_t = f(y_{t-1}, y_{t-2}, \ldots, y_{t-n}) \tag{7}$$

The network used in this paper is shown in Fig. 3.

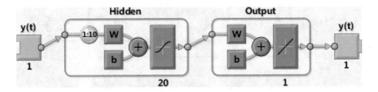

Fig. 3. Adopted NAR network.

Time series are predicted by these two models, on two levels of the whole market and individual stocks. Dow Jones Industrial Average (^DJI), NYSE Composite (^NYA), NASDAQ Composite (^IXIC) and S&P 500 (^GSPC), are the four indices for the whole market. Trading data of 616 listed companies are collected for the level of individual stocks. First, 286 daily trading data, from October 29, 1992 to December 15, 1993, is adopted to calculate logarithm returns and train the networks. The next 21 daily returns named the test window, from December 15, 1993 to January 14, 1994, are predicted, as shown in Fig. 4. All trading data is from http://finance.yahoo.com/.

These expected returns are compared with the observed returns, and the absolute errors are shown in Fig. 5.

For the four indices, the mean absolute errors (MAE) across the test window are 0.0038, 0.0027, 0.0030 and 0.0024 from LSTM, those from NAR network are 0.0033, 0.0030, 0.0030 and 0.0026; the mean squared errors (MSE) from LSTM are 0.0002232, 0.0001478, 0.0001533 and 0.0001151, those from NAR network are 0.0002071, 0.0001656, 0.0001480 and 0.0001178.

For the individual stocks, MAE and MSE of errors across the test window are shown in Fig. 6.

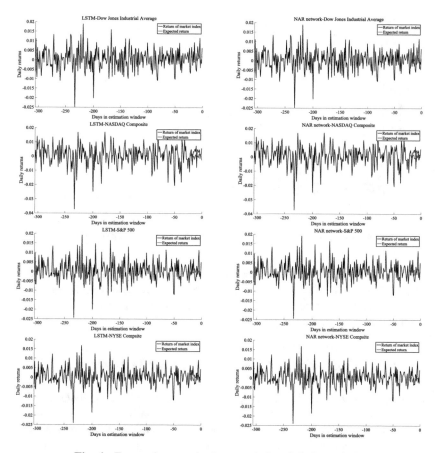

Fig. 4. Expected returns in the test window (whole market).

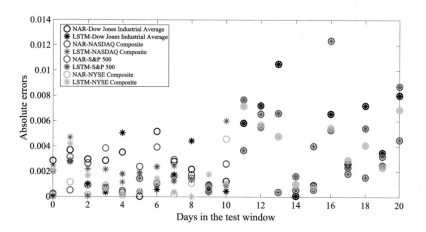

Fig. 5. Absolute errors in the test window (whole market).

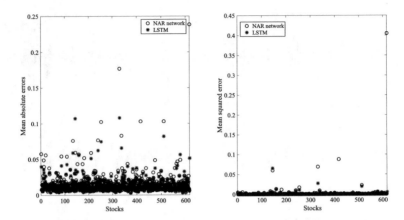

Fig. 6. Comparison of two models in the test window (individual stocks).

From both Figs. 5 and 6, it is not clear which model is more effective. After 10 days, both MAE and MSE from two models are similar, and the increased absolute error indicates that the predictive ability is limited. So, in the following, two models are used for the next 10 trading days.

3   Test the Abnormal Returns in the Event Window

The event window, from January 17, 1994 to January 31, 1994, includes 11 daily returns ($\tau = 0,...,10$). The estimation window is from October 29, 1992 to January 14, 1994, including 306 daily returns ($t = -306,..., -1$). Following the methodology of event study, the abnormal return $AR_{i\tau}$ of market index or stock i on day τ is defined as the residuals between the actual returns $R_{i\tau}$ and the expected ones $E(R_{i\tau})$ from these two models. The abnormal return observations must be aggregated in order to draw overall inferences for the event of interest [21]. Abnormal return is aggregated through the event window and cumulative abnormal return (CAR) at day τ is

$$CAR_{i\tau} = \sum_{\tau=0}^{20} AR_{i\tau} \tag{8}$$

To illustrate the event effect, the significant of abnormal returns needs to be tested. We use two typical test statistics of parameter test and non-parameter test.

Under the null hypothesis H_0, the distribution of standardized $CAR_{i\tau}$ is Student t-test with T-2 degrees of freedom, where T is the length of the estimation window. For a large estimation window ($T > 30$), the distribution will be well approximated by the standard normal [21], the test statistics θ_1 are

$$\theta_1 = \frac{CAR_{i\tau}}{std(CAR_{iT})} \sim N(0,1) \tag{9}$$

where, the standard deviations $std(CAR_{iT})$ are calculated by the CAR_i in the estimation window.

To avoid the discussion on the distribution of abnormal return, nonparametric test is suggested including sign test and rank test [11, 22, 23], which enables one to check the robustness of conclusions based on parametric tests.

Assuming cumulative abnormal returns are independent across securities in the sign test, the null hypothesis is the proportion of stocks with positive cumulative abnormal returns in the event window equals to the expected proportion without abnormal performance. The test statistic θ_2 of the sign test is

$$\theta_2 = \frac{N^+ - N\hat{p}}{\sqrt{N\hat{p}(1-\hat{p})}} \sim N(0,1) \tag{10}$$

where, N^+ is the number of stocks with positive abnormal returns or cumulative abnormal returns in the event window; N is the number of stocks; $\hat{p}$ is the expected proportion of stocks with positive abnormal returns, obtained from the estimation window, $\hat{p} = \frac{1}{N}\sum_{i=1}^{N}\frac{1}{T}\sum_{t=-T}^{-1}S_{it}$; S_{it} depends on the sign of abnormal returns, $S_{it} = 1$ if abnormal return is positive, $S_{it} = 0$ if abnormal return is negative.

For the whole market, the test statistics θ_1, on the event day and the following 10 days, is shown in Fig. 7.

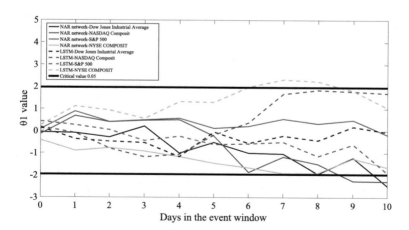

Fig. 7. The test statistics θ_1 of four market indices.

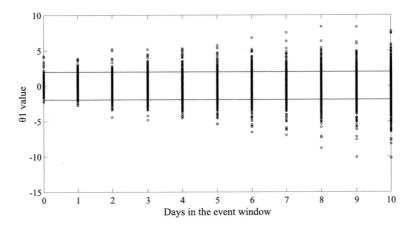

Fig. 8. The test statistics θ_1 of stocks.

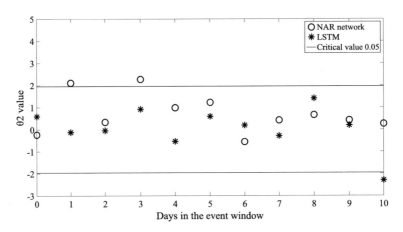

Fig. 9. The test statistics θ_2 of stocks.

In this figure, the black solid line denotes the significant at 0.05 level. It is shown that the null hypothesis H_0 is not rejected from the results of both two models. That is, the reaction of the US stock market is not significant in 10 days after the event.

To detect the reaction of individual stocks, the test statistics θ_1, on the event day and the following 10 days, is shown in Fig. 8. In this figure, circle "o" is for θ_1 from NAR network, star "*" is for θ_1 from LSTM.

The impacted companies in the whole event window, according to the results of both two models, are listed in Table 1.

Table 1. Impacted companies.

Impact	Listed company	Days	Industry
Positive (20)	AAR Corp. (AIR)	10	Aerospace & Defense
	Alaska Air Group, Inc. (ALK)	10	Airlines
	Capstead Mortgage Corp. (CMO)	6–7	REIT - Residential
	CoreLogic, Inc. (CLGX)	10	Business Services
	Corning Inc. (GLW)	10	Electronic Components
	Equity Commonwealth (EQC)	8–10	REIT - Office
	Forest City Enterprises Inc. (FCE-A)	7–8	REIT - Retail
	Lennar Corporation (LEN)	7–8	Residential Construction
	Louisiana-Pacific Corp. (LPX)	10	Building Materials
	Meredith Corporation (MDP)	10	Publishing
	National Presto Industries Inc. (NPK)	10	Aerospace & Defense
	Newell Rubbermaid Inc. (NWL)	10	Household & Personal Products
	Park Electrochemical Corp. (PKE)	8–9	Semiconductor Equipment & Materials
	Pepsico, Inc. (PEP)	8	Beverages - Soft Drinks
	Rogers Corporation (ROG)	10	Electronic Components
	Unifi Inc. (UFI)	10	Textile Manufacturing
	Universal Health Services Inc. (UHS)	9–10	Medical Care
	V.F. Corporation (VFC)	10	Apparel Manufacturing
	Valero Energy Corporation (VLO)	10	Oil & Gas Refining & Marketing
	Watsco Inc. (WSO)	8	Electronics Distribution
Negative (23)	Alexander's Inc. (ALX)	10	REIT - Retail
	BP Prudhoe Bay Royalty Trust (BPT)	4, 7–10	Oil & Gas Refining & Marketing
	BP p.l.c. (BP)	10	Oil & Gas Integrated
	Becton, Dickinson and Company (BDX)	10	Medical Instruments & Supplies
	Carlisle Companies Incorporated (CSL)	10	Conglomerates
	Colgate-Palmolive Co. (CL)	10	Household & Personal Products
	Dycom Industries Inc. (DY)	10	Engineering & Construction
	Entergy Corporation (ETR)	10	Utilities - Diversified
	Ferro Corporation (FOE)	10	Specialty Chemicals
	GATX Corp. (GMT)	9–10	Rental & Leasing Services
	ION Geophysical Corporation (IO)	10	Oil & Gas Equipment & Services
	Polaris Industries Inc. (PII)	7	Recreational Vehicles
	Ramco-Gershenson Properties Trust (RPT)	8	REIT - Retail
	Standard Motor Products Inc. (SMP)	10	Auto Parts
	Stryker Corporation (SYK)	10	Medical Devices
	TEGNA Inc. (TGNA)	10	Broadcasting - TV
	Telephone & Data Systems Inc. (TDS)	10	Telecom Services
	Tutor Perini Corporation (TPC)	10	Engineering & Construction
	UGI Corporation (UGI)	10	Utilities - Regulated Gas
	UnitedHealth Group Incorporated (UNH)	10	Health Care Plans
	Winnebago Industries, Inc. (WGO)	10	Recreational Vehicles
	Wolverine World Wide Inc. (WWW)	10	Footwear & Accessories
	World Fuel Services Corp. (INT)	10	Oil & Gas Refining & Marketing

There are 20 stocks with positive reaction, in which 5 stocks are from housing-related industry and 5 are from manufacturing industry; 23 stocks react negatively, in which 8 stocks are from service industry.

The test statistics θ_2, on the event day and the following 10 days, is shown in Fig. 9.

The result from sign test shows that the stock market is not affected significantly, which is the same with the results from market indices.

4   Conclusions

Two deep learning models, NAR network and LSTM, are employed to predict the normal situation of US stock market. One reason is to test the validation of the models, the other one is to keep the robustness of the results. The result in the test window shows that these two models are equally effective, even though the NAR network is simpler than the typical LSTM. To quantize the impact of the Northridge Earthquake on the US stock market, abnormal returns in the event window are test by a parameter test (T-test) and a non-parameter test (sign test). The result from both two models is that the impact on the whole market is not significant at 95% confidence level, 20 stocks with positive reaction, and 23 stocks react negatively.

Acknowledgement. This work was funded by Scientific Research Fund of Institute of Engineering Mechanics, China Earthquake Administration (Grant No. 2013B14) and National Nature Science Foundation of China (51678540, 51478443 and 51178435).

References

1. Public Policy Institute of California, Northridge Earthquake. https://www.ppic.org/content/pubs/jtf/JTF_NorthridgeJTF.pdf. Accessed 22 Jan 2019
2. USGS, USGS Response to an Urban Earthquake-Northridge '94. https://pubs.usgs.gov/of/1996/ofr-96-0263/. Accessed 21 Jan 2019
3. Risk Management Solutions: The Northridge, California Earthquake, RMS 10-year Retrospective (2004)
4. USGS. https://earthquake.usgs.gov/earthquakes/eventpage/ci3144585/executive. Accessed 22 Jan 2019
5. California Earthquake Authority, Northridge earthquake remembered as one of costliest natural disasters in U.S. history. https://www.earthquakeauthority.com/Press-Room/Press-Releases/2019/Northridge-earthquake-remembered. Accessed 22 Jan 2019
6. Insurance Information Institute, Inc., Background on: Earthquake insurance and risk. https://www.iii.org/article/background-on-earthquake-insurance-and-risk. Accessed 18 Jan 2019
7. Goltz, J.: The Northridge, California Earthquake of January 17, 1994: General Reconnaissance Report, Technical Report NCEER-94-0005. National Center for Earthquake Engineering Research, Buffalo, New York (1994)
8. Tierney, K.: Business impacts of the Northridge Earthquake, Preliminary paper, University of Delaware, Disaster Research Center (1996)

9. Eguchi, R., Goltz, J., Taylor, C., Chang, S., Flores, P., Jphnson, L., Seligson, H., Blais, N.: Direct economic losses in the Northridge Earthquake: a three-year post-event perspective. Earthquake Spectra **14**(2), 245–264 (1998)

10. Aiuppa, T., Krueger, T.: Insurance stock prices following the 1994 Los Angeles earthquake. J. Insur. **18**, 23–36 (1995)

11. Lamb, R., Kennedy, W.: Insurer stock prices and market efficiency around the Los Angeles earthquake. J. Insur. Issues **20**(Spring), 10–24 (1997)

12. Marlett, D., Corbett, R., Pacini, C.: Insurer stock price responses to the disclosure of revised insured loss estimates after the 1994 Northridge Earthquake. J. Insur. Issues **23**(2), 103–123 (2000)

13. Marlon, B.: Business losses, transportation damage and the Northridge Earthquake. J. Transp. Stat. **1**(2), 49–64 (1998)

14. Gordon, P., Richardson, H., Davis, B.: Transport-related impacts of the Northridge Earthquake. J. Transp. Stat. **1**(2), 21–36 (1998)

15. Adam, R., Benavides, J., Chang, S., Szczesniak, P., Lim, D.: The regional economic impact of an earthquake: direct and indirect effects of electricity lifeline disruptions. J. Reg. Sci. **37**(3), 437–458 (1997)

16. Hochreiter, S., Schmidhuber, J.: Long short-term memory. Neural Comput. **9**(8), 1735–1780 (1997)

17. Liu, J., Zhang, T., Han, G., Gou, Y.: TD-LSTM: temporal dependence-based LSTM networks for marine temperature prediction. Sensors **18**(11), 3797 (2018)

18. Yuan, X., Chen, C., Lei, X., Yuan, Y., Adnan, Y.: Monthly runoff forecasting based on LSTM-ALO model. Stoch. Env. Res. Risk Assess. **32**, 2199–2212 (2018)

19. Zhang, R., Huang, C., Zhang, W., Chen, S.: Multi factor stock selection model based on LSTM. Int. J. Econ. Finance **10**(8), 36–42 (2018)

20. Zhang, T., Song, S., Li, S., Ma, L., Pan, S., Han, L.: Research on gas concentration prediction models based on LSTM multidimensional time series. Energies **12**(1), 161 (2019)

21. Campbell, Y., Lo, W., MacKinlay, C.: The Econometrics of Financial Markets. Princeton University Press, New Jersey (1998)

22. Javid, Y.: Stock market reaction to catastrophic shock-evidence from listed Pakistani firm. Working Papers, Pakistan Institute of Development Economics (2007)

23. Scholtens, B., Voorhorst, Y.: The impact of earthquakes on the domestic stock market. Earthquake Spectra **29**(1), 325–337 (2013)

SWT-ARMA Modeling of Shenzhen A-Share Highest Composite Stock Price Index

Jingyi Wu[✉]

School of Management Science and Engineering,
Central University of Finance and Economics, Beijing 10226, China
h08412lwjy@126.com

Abstract. Simulation and prediction of stock trading price time series have always been one of the important research contents in the financial investment industry. Based on the data from the national stock trading statistics table, SWT and ARMA steady-state modeling, the multiscale simulation and prediction model SWT-ARMA is constructed on the time series of the highest composite stock price index Shenzhen A-share, which quantitative simulation analyzes and embedded predicts the evolutionary trend of the sequence and its effectiveness. The results shows that the fitting ability and extrapolation prediction accuracy of the multiscale SWT-ARMA model are higher than the accuracy of the ARMA model, and it is adaptive to the smoothing preprocessing and modeling of the stock data. Therefore, this method can effectively model and predict the time series of stock trading prices.

Keywords: Shenzhen · The highest composite stock price index of a shares · Multiscale SWT-ARMA model

1 Introduction

The simulation and prediction of time series variation characteristics of stock trading has always been one of the most important research contents of the financial investment securities industry. Its evolution trend and development process are complex and generally affected by a variety of factors. The dynamic and non-linear phenomena do not generally meet the requirements of stability. In order to finely simulate and characterize the complexity of sequences, nonlinear methods or artificial intelligence are generally used in recent years [1–4]. These methods improve the accuracy of sequence modeling, but also solidify the dependence between model and data, increasing the complexity of modeling and reducing the efficiency of data processing.

In order to overcome the shortcomings of the above methods, it is necessary to take advantage of the classical methods, such as AR, ARMA (Auto-Regression and Moving Average) [5–11]. However, these method requires the sequence to be stable [6, 7], while the actual stock trading sequence generally does not meet this assumption. This makes it difficult for sequence modeling and steady prediction. In order to fully utilize and dig out the inherent advantages of traditional methods, it is generally required to first smooth and linearize the actual sequence, and then model and predict it.

© Springer Nature Switzerland AG 2020
J. H. Kim et al. (Eds.): ICHSA 2019, AISC 1063, pp. 122–129, 2020.
https://doi.org/10.1007/978-3-030-31967-0_14

There are many pretreatment methods available for selection in the stabilization process of unsteady time series [9, 11]. Different pretreatments produce different stationary sequences, which will complicate the modeling process and inevitably lead to the randomness, multi-solution and non-interpretability of the modeling and prediction results.

In fact, with the rapid development of modern signal processing technology, many effective methods of smoothing have been produced, and SWT (Stable wavelet transform) is one of them [12]. Research shows that SWT is an effective signal frequency division processing tool after wavelet transform. It has multi-resolution characteristics and translation invariance, and its results have certain physical meaning and interpretability, which provides the guarantee of smoothing data for the analysis of related problems. For the smoothing of non-stationary sequences with dynamic and nonlinear changes, SWT technology is naturally a suitable choice.

Therefore, SWT [12] technology is used to obtain stable high-frequency and low-frequency sequences, and ARMA method with relatively simple method principle and strong fitting ability is further selected [5, 8–11]. In this way, the SWT-ARMA multiscale model for the highest composite stock price index of Shenzhen A-share is established, and quantitative simulation and prediction tests are carried out.

2 Multiscale Simulation and Prediction Based on SWT and ARMA Model

2.1 Basic Principle of SWT Positive and Inverse Transformation

In this paper, A Trous algorithm is adopted, which is mainly featured by its simplicity, fast speed and small amount of calculation [12]. The SWT mainly includes positive and inverse transformation. Its basic principle is introduced as follows:

A scale function $\phi(t)$ and a wavelet function $\varphi(t)$ are set, and the scale function $\phi(t)$ satisfies the two-scale equation:

$$\frac{1}{2}\phi\left(\frac{t}{2}\right) = \sum_{n=-\infty}^{+\infty} h(n)\phi(t-n) \tag{1}$$

$h(n)$ is the coefficient of discrete low-pass filter.

The definition of $C_0(t)$ is that the approximate sequence of 0 scales is obtained by the inner product of signal $f(t)$ and scale $\phi(t)$ function.

$$C_0(t) = <f(t), \phi(t)> \tag{2}$$

In practical applications, it is often taken $C_0(t) = f(t)$ as the initial value. Then the approximate series (low-pass series) at scale $j-1$ is the inner product of the approximate sequence and scale at scale:

$$C_j(t) = <C_{j-1}(t), \phi(t)> \tag{3}$$

From formula (1) and formula (3), we can get:

$$C_j(t) = \sum_{n=-\infty}^{+\infty} h(n)C_{j-1}(t + 2^j n) \tag{4}$$

Since the relation between wavelet function $\varphi(t)$ and scaling function $\phi(t)$ is:

$$\frac{1}{2}\varphi\left(\frac{t}{2}\right) = \phi(t) - \frac{1}{2}\phi\left(\frac{t}{2}\right) \tag{5}$$

Then the detail sequence $W_j(t)$ (high-frequency sequence) at scale j is the difference information between the approximate sequence (low-frequency sequence) at scale j and scale $j - 1$.

$$W_j(t) = C_{j-1}(t) - C_j(t) \tag{6}$$

where, $\{W_1(t), W_2(t), \ldots, W_J(t), C_J(t)\}$ is called wavelet transform sequence in scale J. J is the maximum decomposition scale number. To determine J, there is no definite method at present. Generally speaking, there are at most $\lg N$ (N is the sequence length) scales. This is the A Trous decomposition algorithm.

A Trous reconstruction algorithm is very simple, that is, the sum of the high-frequency sequence of each layer and the low-frequency sequence of the last decomposition layer.

$$C_0(t) = C_J(t) + \sum_{j=1}^{J} W_j(t) \tag{7}$$

2.2 Multiscale Modelling

In formula (7), sequences $W_j(t)$ of each dimension and the largest scale of $C_J(t)$ decomposed by the SWT has a certain degree of stationarity, and sequence $W_j(t)$ itself also satisfies orthogonality relationship, which can completely reconstruct decomposition sequence. Compared with the original sequence, its error is very small, which is the basis of multiscale model based on SWT.

Based on each scale $W_j(t)$ and final scale sequence $C_J(t)$, ARMA model can be established [5, 8–11] to obtain the parameters of ARMA autoregression and moving coefficient of sequence $W_j(t)$ and $C_J(t)$, and then extrapolate the prediction using AMRA. Furthermore, the ARMA results of multiscale sequence $W_j(t)$ and sequence $C_J(t)$ are reconstructed and synthesized according to formula (7). Finally, the original non-stationary time series are modeled and predicted by steady-state method. For the convenience of description, this paper calls it as SWT-ARMA model. The core idea and specific process of multiscale modeling are expressed in Fig. 1.

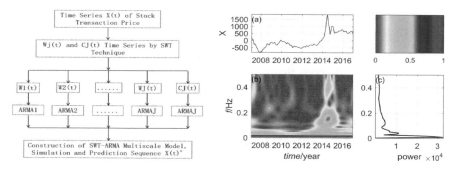

Fig. 1. ARMA modeling process of stock trading time series based on SWT.

Fig. 2. Time series of Shenzhen A-share highest composite stock price index and its S transform feature.

3   Time Series Modeling and Testing of Shenzhen A-Share Highest Composite Stock Price Index

3.1   Time Series Data Selection and Stationarity Test of Shenzhen A-Share Highest Composite Stock Price Index

The time series data of Shenzhen A-share highest composite stock price index (Fig. 2a) selected in the study are from the national stock trading statistics table provided by Dongfang Fortune website (http://data.eastmoney.com/cjsj/gpjytj.html), which is from January 2008 to December 2017 and the sampling unit is month. For the sequence data, the following preliminary analysis is made:

Firstly, it can be seen from the original sequence that the change process of the sequence is not stable and has obvious nonlinear dynamic evolution characteristics. It is initially confirmed from the time domain perspective that it does not conform to a stable time series. Then, based on the S transform [13], time-frequency estimation is performed for the sequence (Fig. 2). According to the time-frequency calculation results in Fig. 2, it is not difficult to find that the frequency changes non-linearly with time, and the energy is not evenly distributed in the time-frequency domain, and there is clustering phenomenon, including trend, period and random noise. From the perspective of time-frequency domain, it is further confirmed that it has non-stationary characteristics.

Both above original curves and time-frequency distribution results show that the time series of stock prices are non-stationary and non-linear. Therefore, before modeling by using the steady-state method, $W_j(t)$ and $C_J(t)$ time series should be obtained through the SWT introduced above to provide steady-state conditions for multiscale modeling.

3.2   SWT and ARMA Modeling of Shenzhen A-Share Highest Composite Stock Price Index Time Series

Multiscale simulation and prediction of the SWT-ARMA model is carried out using the highest composite stock index time series of Shenzhen A-share (Fig. 3a). Among the 120-month data of the original time series represented in Fig. 3a, the first 100-month data are used for multiscale decomposition and simulation, and the last 20-month data are used for the attached prediction test to evaluate the validity of the simulation time series and the robustness of the prediction results.

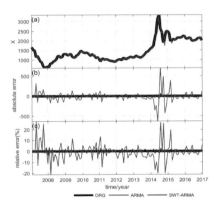

Fig. 3. Original sequence, SWT-ARMA and ARMA model simulation, prediction, residual diagram of Shenzhen A-share highest composite stock price index.

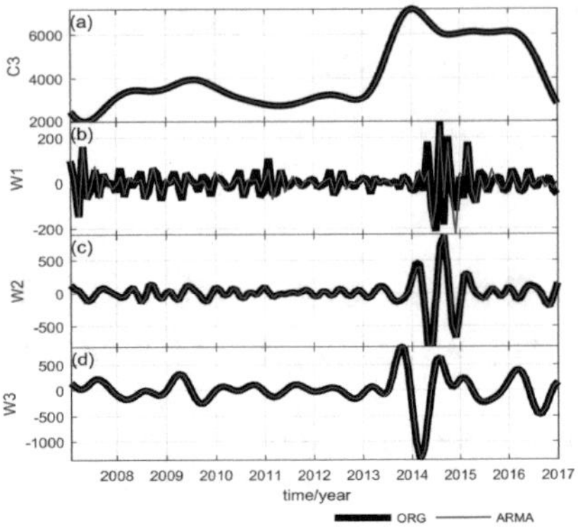

Fig. 4. SWT and ARMA model simulation and prediction results of the highest composite stock price index of Shenzhen A-share.

First of all, SWT is performed on the data of the first 100 months of the stock sequence to obtain the low-frequency and high-frequency sequence results of the sequence, namely, the stationary time series of $W_j(t)$ and $C_J(t)$ (Fig. 4). Then the multiscale sequence $W_j(t)$ and $C_J(t)$ are simulated quantitatively by ARMA, and the autoregressive and moving coefficients of the multiscale ARMA model are obtained. Furthermore, the multiscale prediction of the sequence is carried out with the calculated coefficients. Finally, the prediction sequence is synthesized according to formula (7), and the indirect simulation and comprehensive prediction of the original sequence are finally completed. The simulation, prediction and multiscale intermediate results are shown in Figs. 3 and 4, and the evaluation results in Table 1. From the simulation and prediction results expressed in Figs. 3, 4 and Table 1, the following preliminary understanding can be obtained:

(1) In the time series simulation part of the highest composite stock price index time sequence of Shenzhen A-share, the numerical distribution of each order of sequence $W_j(t)$ and $C_J(t)$ obtained by SWT decomposition of the original sequence is basically symmetrical, and its dynamic property is weakened, its stable property is enhanced. Furthermore, its stationarity obtained by testing [5, 8–11] also meets the stationary

requirements of traditional modeling methods, providing reliable basic data for for multiscale modeling of ARMA method.

(2) ARMA fitting of sequence $W_j(t)$ and sequence $C_J(t)$ shows that the total trend of the simulation results is better and better as the frequency decreases. The simulation of the sequences has certain differences and oscillations within the allowable range only in the high frequency part. Based on the obtained multiscale autoregressive and moving coefficients, the multiscale prediction of the sequence is carried out. Although the prediction accuracy decreases with the extrapolation delay, the prediction accuracy is quite high on the whole, which conforms to the general rule of extrapolation prediction.

(3) Comprehensive comparison of simulation and prediction results (Table 1) shows that the sequence development trend and dynamic evolution details are consistent and synchronous with the known data. The correlation is strong and the error is small (Table 1), indicating that the fitting and prediction accuracy of the sequence is high and the results are reliable.

3.3 Comprehensive Comparison Between SWT-ARMA and ARMA Models

The calculation results of the simulation and prediction results of sequences modeling by SWT-ARMA and ARMA are compared by correlation coefficient, normalized variance, and their normalized comprehensive indexes (Table 1). It is not difficult to see from the comparative analysis of Fig. 3 and Table 1 that:

(1) The correlation coefficient and normalized variance of the simulated sequence and the original sequence obtained by the SWT-ARMA model are both higher than the analysis results of the ARMA model, and the predicted results are going to be the same. The advantages of the SWT-ARMA model are even more so by using the comprehensive evaluation of normalized variance and correlation coefficient. All results of the simulation and prediction sequences are evaluated together and the same results are obtained. Further analysis shows that the SWT-ARMA model can accurately simulate and predict the development trend and detailed changes of sequences, such as the consistency of sequence trend, the periodicity of fluctuations, and the accuracy of poles.

Table 1. Evaluation table of SWT-ARMA and ARMA model simulation and prediction.

Compare content	Coefficient of association	Normalized variance	Variance + correlation coefficient
Entire SWT-ARMA	0.999502	0.015794	0.016039
Entire ARMA	0.998210	0.029932	0.030801
SWT-ARMA simulation	0.999445	0.016720	0.016992
SWT-ARMA prediction	0.999742	0.011406	0.011533
ARMA simulation	0.997991	0.031729	0.032702
ARMA prediction	0.999174	0.020396	0.020800

(2) Compared with ARMA, the accuracy of simulation and prediction of SWT-ARMA model has been greatly improved, and the decline of prediction accuracy is relatively slow. This shows that SWT-ARMA model has a strong generalization ability for the distribution of sequence data, and it is efficient to mine the implicit information inside the sequence. The reason may be that the non-stationary series become random, trend and periodic components after SWT decomposition, so that the steady-state of single-frequency series of $W_j(t)$ and $C_J(t)$ is enhanced, the randomness is weakened, and the mutability is minimized, which is more conducive to the steady-state gradient method for modeling, and ultimately can improve the robustness and adaptability of the model.

Through the above comprehensive comparative analysis of SWT-ARMA and ARMA models from different perspectives, the effectiveness of SWT-ARMA multiscale model in the processing of Shenzhen A-share highest composite stock index sequence is verified and evaluated. Therefore, this model can be used for the modeling and prediction of stock sequences, and the change level and development trend of stock trading prices can be grasped in real time.

4   Discussion and Conclusion

(1) SWT is a kind of stationary frequency division technology, which can transform complex sequences into relatively simple stationary sequences. The decomposition process is objective and adaptive. This is also well proved in the actual decomposition process of Shenzhen A-share highest composite stock price index sequence. According to the SWT definition and decomposition process, it is not difficult to see that the obtained sequence has basically met the inherent requirements of stationary processing, while the decomposition results sequence and sequence have practical physical significance, which is more conducive to the explanation and in-depth analysis of related problems.

(2) The SWT technique is used to preprocess the original stock time series to obtain a multiscale steady-state sequence. It is more advantageous than simple differential methods because it can exploit more "useful" information hidden inside the stock sequence. The multiscale steady-state model is established on the basis of sequence and sequence, and the multiscale simulation and prediction results are synthesized. Compared with the traditional modeling method, the results can significantly improve the modeling ability and prediction accuracy of stock sequences.

Summarizing the above analysis, in the actual processing of stock data, the SWT-ARMA model has more advantages than the ARMA model. Therefore, the SWT technology can be used to transform the dynamic and nonlinear stock time series, and then the traditional and effective steady-state modeling method can be used to obtain a more accurate multiscale model. Finally, a relatively simple steady-state model is used to simulate and predict complex time series.

References

1. Long, J.C., Li, X.P.: Study of the stock market tendency based on the neural network. J. Xidian Univ. **32**(3), 460–463 (2005)
2. Xin, Z.Y., Gu, M.: Complicated financial data time series forecasting analysis based on least square support vector machine. J. Tsinghua Univ. (Sci. Technol.) **48**(7), 1147–1149 (2008)
3. Wu, T., Yu, S.W.: Quantitative Analysis of Stocks Based on Machine Learning. Machinery Industry Press, Beijing (2015)
4. Wang, Z.D.: Quantitative Investment Expert System Development and Strategy Practice. Electronic Industry Press, Beijing (2018)
5. Quan, F.S., Peng, B.Y.: ARMA model application in China's stock market. J. Hengyang Normal Univ. **30**(3), 26–28 (2009)
6. Tsay, R.S.: Analysis of Financial Time Series, 3rd edn. Wiley, New Jersey (2010)
7. Shumway, R.H., Stoffer, D.S.: Time Series Analysis and its Applications: with R Examples, 3rd edn. Springer, New York (2011)
8. Li, L.: Research on the dynamic relationship of stock market quantitative prices based on the ARMA-GARCH model. Stat. Decis.-Making **4**, 144–146 (2011)
9. Li, Y., Zheng, Z.Y.: Quantitative Investments: MATLAB as a Tool. Beijing Electronic Industry Press, Beijing (2015)
10. Yang, Q., Cao, X.B.: Analysis and prediction of stock price based on ARMA-GARCH model. Math. Pract. Theory **46**(6), 80–86 (2016)
11. Huang, Y.R.: Quantitative Analysis of Finance. Tsinghua University Press, Beijing (2018)
12. Gao, Z., Yu, X.H.: Matlab Wavelet Analysis and Application, 2nd edn. National Defense Industry Press, Beijing (2007)
13. Stoekwell, R.G., Mansinha, L., Lowe, R.P.: Localization of the complex spectrum: the S-transform. IEEE Trans. Signal Process. **17**(6), 998–1001 (1996)

The Application of Text Categorization Technology in Adaptive Learning System for Interpretation of Figures

Weibo Huang[✉], Zhenpeng He, and Xiaodan Li

Guangdong University of Foreign Studies,
Guangzhou 510006, Guangdong, People's Republic of China
172564946@qq.com

Abstract. With the deepening of globalization and the increasing amount of information in international communication, requirements for accuracy in interpretation of figures are more demanding. Training methods in interpretation of figures, however, are not efficient enough for interpreters to cope with the challenges. They still make mistakes in interpretation of large integers, fractions and percentages. The types of errors include omission, syntactic error and lexical error. In this paper, machine learning based text categorization technology is used to accurately categorize a large number of texts and provide high-quality training materials for interpreters. Results show that training the interpreters with categorized texts has greatly improved the accuracy, familiarity and sensitivity in interpretation of figures. In the era of artificial intelligence, problems in interpretation also need to be solved by artificial intelligence. In the future, a large number of artificial intelligence technologies similar to machine-learning-based text categorization technology will be inevitably adopted in the field of interpretation.

Keywords: Machine learning · Text categorization · Interpretation · Interpretation of figures · Adaptive learning system

1 Introduction

In simultaneous interpretation, a large number of figures have to be accurately interpreted into the target language in order to accurately convey the information in the text [1]. Thus, interpreters will be under great pressure in interpretation of texts with a great number of figures [2]. Therefore, interpreters should have a larger amount of effective training in interpretation of figures [3]. Current training methods in interpretation of figures are mainly manual methods. Machine learning is the core of artificial intelligence and data science. Machine learning is the intersection of computer science and statistics, aiming at improving the performance of the system by experience and is widely applied in various research fields, such as fraud detection, fault diagnosis, speech recognition, chess, machine translation, and image search system in Google [4].

© Springer Nature Switzerland AG 2020
J. H. Kim et al. (Eds.): ICHSA 2019, AISC 1063, pp. 130–138, 2020.
https://doi.org/10.1007/978-3-030-31967-0_15

On the basis of current research at home and abroad, this paper uses machine learning based text classification technology to classify unclassified text data (such as government work reports, speeches, magazine articles, etc.), extract sentences with a lot of figures for interpreters to practice, analyze and summarize, and design better training method in interpretation of figures for the adaptive learning system [4].

2 Current Studies on Machine Learning Classification Technology

Studies on machine learning classification technology mainly focuses on the following aspects:

The Positive Correlation. In the methods of text classification in machine learning, there is a positive correlation between the accuracy of classification and the difficulty in the design of the system. So word embedding, deep learning algorithm and multi-model fusion are proposed to improve the text classification system [5].

The Bidirectional User Interests' Set. Based on the traditional machine algorithm, the bidirectional user interests' set is introduced to solve the problems of spam filtering, low matching degree and insufficient type of personalized news recommendation [6].

The Neural Network Model. The neural network model is used to conduct experiments on features such as emotion, location and part of speech, so as to classify social network texts and extract information [7].

At present, the application of machine learning text classification technology is still mainly used in the identification of bad advertisements, website titles, hot topics on social networks, advertising recommendations on the Internet and other aspects [6]. However, little research is about the improvement of the interpretation of figures. Based on machine learning text classification technology, this study proposes possible solutions to existing problems in the interpretation of figures.

3 Application of Machine Learning Text Categorization Technology

According to the above subjective and objective factors and the limitations of current training methods, this paper adopts machine-learning-based text categorization technology to improve the accuracy in interpretation of figures.

This research adopts Bag of Words. A simple example illustrates how One-hot in Bag of Words extracts text characteristic vectors. Assuming that there are three sentences in the corpus: "give priority to recommending high-quality enterprises with high-tech features", "give priority to recommending high-quality enterprises" and "recommend high-quality enterprises with high-tech features" [8]. The corpus is separated and all the words are obtained. Then each word is coded: "Give priority to" is coded as 1, "recommend" is coded as 2, "with high-tech features" is coded as 3, and "high-quality enterprises" is coded as 4. One-hot is used to extract feature vectors from each paragraph, as shown in Fig. 1.

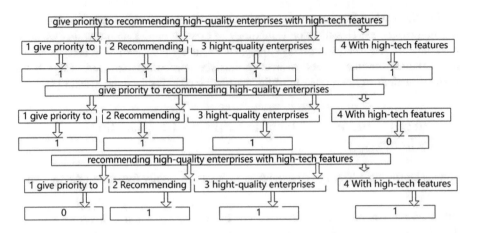

Fig. 1. Example of BOW.

Therefore, feature vectors of "give priority to recommending high-quality enterprises with high-tech features", "give priority to recommending high-quality enterprises" and "recommend high-quality enterprises with high-tech features" are $\{1, 1, 1, 1\}$, $\{1, 1, 1, 0\}$, $\{0, 1, 1, 1\}$ respectively. The advantage of One-hot is that it solves the problem of weak ability of classifier to process discrete data and extends the characteristics of text. This paper adopts Bag of Words to extract the characteristics of the text.

Characteristic Extraction. After pre-processing and text representation, there still exist words unrelated to the topic of the text. The computer, however, focuses on the words related to the topic, so the research needs to add weight to the corresponding words. In the research, TF-IDF algorithm will be used to add weight to the characteristics [9].

TF is the frequency of a word in a text, that is, word frequency.

$$tf_{i,j} = \frac{n_{i,j}}{\sum_k n_{i,j}} \tag{1}$$

IDF is the frequency of reversed-document: the equation is:

$$idf_i = \log \frac{|D|}{|\{j : t_i \in d_i\}|} \tag{2}$$

$|D|$ represents the total number of documents in the training set; $\{j: t\_dj\}$ | represents all documents containing the word Ti. Generally, $\{j: t\_dj\}$ | may be 0 (that is, the number of documents containing the word is 0), so the denominator is +1:

$$1 + |\{j : t_i \in d_j\}| \tag{3}$$

Equation (2) and (3) divides the number of documents containing this word by the total number of documents and then logarithms them. Specifically, if there are very few documents containing a specific word, this word will be very distinguishing ("of" in most articles, "index" in most articles about search engines, so the word "index" is more distinguishing than "of").

TF/IDF method is to combine TF with IDF. The greater the value of TF * IDF is, the greater the weight of the word will be. Then this word is more important to the document. When the more frequently a word appears in a document, while the less frequently it appears in other documents, then this word will have higher weight.

Official Definition. Chi-square test is a common hypothesis test method based on X^2 distribution. H_0, an invalid hypothesis, is that there is no difference between the observed frequency and the expected frequency. The basic idea of this test is to assume that H_0 is valid. Based on this assumption, the value of X_2 is calculated, which shows the degree of deviation between the observed value and the theoretical value. According to the distribution and degree of freedom of X2, the probability P of current statistics and more extreme cases can be calculated under the assumption of the validity of H_0. If the current statistic is larger than P value, the deviation between the observed value and the theoretical value is too large. It indicates that comparative data is significantly different, so assumption H_0 should be rejected; otherwise, assumption H_0 should not be rejected and it does not indicate that the actual situation of the sample is different from the theoretical one.

The Basic Equation of Chi-square Test.

$$\chi^2 = \sum \frac{(A - E)^2}{E} = \sum_{i=1}^{K} \frac{(A_I - E_i)^2}{E_i} = \sum_{i=1}^{k} \frac{(A_i - np_i)^2}{np_i} \tag{4}$$

Equation deduction:

X^2 is the deviation between the observed value and the theoretical value. A is the observed value (observed frequency), E is the theoretical value (that is, the value assumed), K is the number of observed value, n is the total frequency, p is the theoretical frequency, and n * p is the theoretical frequency (theoretical value).

The first step is to calculate the deviation between the observed value and the theoretical value, that is, the difference value between them. Sum the deviations between the observed value and the theoretical value.

$$\chi^2 = \sum_{i=1}^{k} (A_i - E_i) \tag{5}$$

As there are many observational and theoretical value, they can be both positive and negative. Once the offset of the result is 0, there will be no original deviation after calculation. Therefore, adjustments need to be made in the second step. The second step is to square the deviation between the observed value and the theoretical value.

The third step is to square, divide by the theoretical value and then sum them up. This will not affect the calculation of the deviation because of the theoretical value.

$$\chi^2 = \sum_{i=1}^{K} \frac{(A_I - E_i)^2}{E_i} \tag{6}$$

The final equation for calculating χ^2 is now obtained.

$$\chi^2 = \sum \frac{(A - E)^2}{E} = \sum_{i=1}^{K} \frac{(A_I - E_i)^2}{E_i} = \sum_{i=1}^{k} \frac{(A_i - np_i)^2}{np_i} \tag{7}$$

Chi-Square Distribution. It can be seen that when the observed value is very close to the theoretical value, that is, when the hypothesis is correct, the value of X^2 tends to be zero. That is to say, the smaller the deviation is, the more likely the hypothesis is correct; while the larger the deviation is, the more likely the hypothesis is inaccurate. The Chi-square distribution is used to test deviation value.

4 Case Analysis

4.1 Equation to Be Used

Chi-Square Test. Make a hypothesis in advance and calculate the deviation between the observed value and the theoretical value to see whether the hypothesis is valid or not. Formula (deduced above):

$$\chi^2 = \sum_{i=1}^{K} \frac{(A_I - E_i)^2}{E_i} \tag{8}$$

The Chi-square test equation of four-grid tables. Chi-square test for text categorization: use a four-grid table. Each word corresponds to each category. Then calculate whether this word has a greater impact on this category.

$$\chi^2 = n \cdot \frac{(AD - BC)^2}{[(A+B)(C+D)(A+C)(B+D)\}} \tag{9}$$

This equation can be deduced by a simple example:

Assume that a series of news headlines are given, whether some words in the news are related to the category of the news can be figured out, as shown in Table 1.

Table 1. Take the categories of news with headlines "Sino-US trade" as an example

Group	International news	Non-international news	Total
Excluding "Sino-US trade"	19	24	43R1
Including "Sino-US trade"	34	10	44R2
Total	53C1	34C2	87

C1, C2, R1, R2 represents the sum of Row 1, Row 2, Line 1 and Line 2 respectively.

The observed value is the figures in the table, which represents the actual value. Obviously, there are four observed value, 19, 24, 34 and 10 in the table.

According to independence test and cross-tab table, equation $K^2 = n \cdot \frac{(AD-BC)^2}{[(A+B)(C+D)(A+C)(B+D)]}$ can be used to calculate the deviation between theoretical value and actual value. If the two variables are independent, the difference between theoretical value and actual value in the four-grid table will be very small. So the difference between theoretical value and actual value can be calculated by the following equation:

$$\chi^2 = n \cdot \frac{(AD - BC)^2}{[(A+B)(C+D)(A+C)(B+D)\}} \tag{10}$$

The value of X2 is 10.00.

The application of X2:

Degree of freedom (df) = (the number of lines −1) * (the number of rows −1), indicates that when calculating a statistic, the value of it is not restricted by the number of variables, as shown in Table 2.

Table 2. X2 distribution chart.

K\p	0.95	0.90	0.80	0.70	0.50	0.30	0.20	0.10	0.05	0.01	0.001
1	0.004	0.02	0.06	0.15	0.45	1.07	1.64	2.71	3.84	6.64	10.83
2	0.10	0.21	0.45	0.71	1.39	2.41	3.22	4.60	5.99	9.21	13.82
3	0.35	0.58	1.01	1.42	2.37	3.66	4.64	6.25	7.82	11.34	16.27
4	0.71	1.06	1.65	2.20	3.36	4.88	5.09	7.78	9.40	13.28	18.47
5	1.14	1.61	2.34	3.00	4.35	6.06	7.29	9.24	11.07	15.09	20.52
6	1.63	2.20	3.07	3.83	5.35	7.23	8.56	10.64	12.59	16.81	22.46
7	2.17	2.83	3.82	4.67	6.35	8.38	9.80	12.02	14.07	18.48	24.32
8	2.73	3.49	4.59	5.53	7.34	9.52	11.03	13.36	15.51	20.09	26.13
9	3.32	4.17	5.38	6.39	8.34	10.66	12.24	14.68	16.92	21.67	27.88
10	3.94	4.86	6.18	7.27	9.34	11.78	13.44	15.99	18.31	23.21	29.59
......											

In general, P is equal to 0.05. That is to say, when the unrelated probability is 0.05, the corresponding chi-square value is 3.84. Obviously, 10.0 > 3.84. It indicates that the probability that news containing "Sino-US trade" does not belong to international news is less than 0.05. In other words, the probability that the news containing "Sino-US trade" is related to international news is more than 95%.

4.2   The Experimental Results (Take Chinese-English Translation as an Example)

Current translation strategies include alphanumeric combination, international general law, tables, three dots multiple four bars, Arabic numeral counting method, decimal point method and so on. But many of them are overlapped. So, generally these methods can be divided into Table and Abbreviation.

This study used these two methods and Adaptive System Correction Method to conduct experiments on 267 undergraduates in Grade 2016 and Grade 2017 in Gangdong University of Foreign Studies.

Select 469 historical articles from NLPIR and 420 articles from a computer technology website.

Firstly, select a word segmentation tool to segment the training set.

The chosen word segmentation tool is "shuttering participle" software, and the language used is C#.

Secondly, calculate word frequency by using stop list.

After Chi-square test, these words in the history category are more frequent: "lishi" (history) (15532 times), "zhong" (middle) (11477 times), "zhongguo" (China) (9434 times), "nian" (year) (7942 times), "fazhan" (development) (5862 times), "yanjiu" (research) (5656 times), "wenhua" (culture) (4737 times), "shehui" (society) (4715 times), "shuo" (say) (3738 times), "xin" (new) (3706 times). The words in the computer category are more frequent: | | (2089 times), "zhong" (middle) (1332 times), {(1288 times), "wenjian" (documents) (819 times), "name" (521 times), "the" (487 times), "lei" (class) (475 times), "shi" (time) (461 times), "fangfa" (method) (456 times), "zhixing" (execution) (417 times).

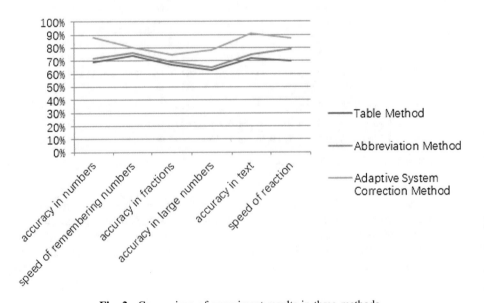

Fig. 2. Comparison of experiment results in three methods

Thirdly, categorize according to the extracted text characteristics. It can be done by statistical categorization methods, such as Nave Bayes, KNN, SVM, maximum entropy, neural network, etc.

Fourthly, use the method mentioned above to select the text with many figures. Provide simultaneous interpretation training for interpreters through audio reading software.

Experiment results were analyzed through Baidu statistics software, as shown in Fig. 2.

Experiment results show that the accuracy in interpretation of figures with the means of Table and Abbreviation is similar. Adaptive system correction method, however, can significantly improve the accuracy in interpretation of figures, but still there is room for improvement in the accuracy in the interpretation of fractions.

5   Summary

More intelligent and automatic methods are needed in the era of artificial intelligence. In this study, machine learning technology was used to classify text. Machine-learning-based text categorization technology in adaptive learning system was used to accurately categorize a large number of texts and provide high-quality training materials for interpreters. The adaptive system can effectively reduce the input of human resources while give interpreters more high-quality training at the same time. It can help interpreters have more efficient preparations before interpretation and build up confidence in interpretation of figures.

Acknowledgement. This study was financially supported by the Undergraduate Innovation Training Project of Guangdong University of Foreign Studies in 2019 (NO. 201911846007).

References

1. Zhang, W., et al.: Research on software defect prediction method based on machine learning. Appl. Mech. Mater. **4**, 687–691 (2014)
2. Petukhov, A.Y., Polevaya, S.A.: Dynamics of information images in the mind of an individual during simultaneous interpretation. Procedia Comput. Sci. **123**, 354–359 (2018)
3. Batic, J., Haramija, D.: The importance of visual reading for the interpretation of a literary text. CEPS J. Center Educ. Policy Stud. J. **5**, 31–49 (2015)
4. Díaz-Galaz, S., Padilla, P., Teresa Bajo, M.: The role of advance preparation in simultaneous interpreting: a comparison of professional interpreters and interpreting students. Interpreting **17**, 1–8 (2015)
5. Salles, T., Rocha, L., Mourão, F., Gonçalves, M., Viegas, F., Meira Jr., W.: A Two-stage machine learning approach for temporally-robust text classification. Inf. Syst. **69** (2017). S0306437917301801
6. Deng, L., Jia, Y., Zhou, B., Huang, J., Han, Y.: User interest mining via tags and bidirectional interactions on Sina Weibo. World Wide Web **21**, 1–22 (2017)

7. Higgins, I., Stringer, S., Schnupp, J.: Unsupervised learning of temporal features for word categorization in a spiking neural network model of the auditory brain. PLoS ONE **12**, e0180174 (2017)
8. Liu, J.-H., Liang, W.-X., Li, X.-D.: Formative assessment system of college english based on the big data. In: Xiong, N., Xiao, Z., Tong, Z., Du, J., Wang, L., Li, M. (eds.) Advances in Computation Science and Computing, vol. 877, pp. 472–480 (2018)
9. Richard, A., Gall, J.: A bag-of-words equivalent recurrent neural network for action recognition. Comput. Vis. Image Underst. **156**, 79–91 (2017)

A Method for Extracting and Simplifying the Stray Capacitance Matrix of the Dry-Type Smoothing Reactor

Tingting Li[1,2(✉)], Shaoyan Gong[1,2], Feng Ji[1,2], Chong Gao[1,2], and Jianhui Zhou[1,2]

[1] State Key Laboratory of Advanced Power Transmission Technology
(Global Energy Interconnection Research Institute Co., Ltd.),
Changping District, Beijing 102209, China
lmmziy@126.com, sygong616@163.com,
jameskeating@163.com, chong820515@163.com,
yzhou130@163.com
[2] Beijing Key Laboratory of High-Power Power Electronics
(Global Energy Interconnection Research Institute Co., Ltd.),
Changping District, Beijing 102209, China

Abstract. This paper mainly presents the extraction and simplification of the end-to-end stray capacitance matrix based on the dry-type smoothing reactor. The method of extracting and simplifying the capacitance matrix of a reactor is introduced in detail through an example. The capacitance matrix is extracted by finite element analysis software and is simplified to an end-to-end stray capacitance C. Two kinds of lightning overvoltage analysis model of three reactors in series are established respectively in the transient analysis software and the simulation results verify the correctness of the extraction and simplification method. Through this method, the end-to-end stray capacitance matrix of a DC dry smoothing reactor prototype is obtained, which is independently designed by our institute and will be applied to the Ultra High Voltage Direct Current (UHVDC) project. The stray capacitance of the reactor is measured by an impedance analyzer and the simulated results are in agreement with the experimental ones, which again show that the method is proper.

Keywords: Extraction and simplification · Stray capacitance matrix · Overvoltage analysis model · Prototype

This thesis is supported by the National Key Research and Development Program "Development and Application of the Anode Saturated Reactor and High Frequency Transformer" (Project No. 2017YFB0903905) of "Key Technology of Medium and High Frequency Magnetic Components for High-power Power Electronic Equipment" (No. 2017YFB0903900).

J. H. Kim et al. (Eds.): ICHSA 2019, AISC 1063, pp. 139–147, 2020.
https://doi.org/10.1007/978-3-030-31967-0_16

1 Introduction

As one of the most important device in UHVDC transmission projects, the dry-type smoothing reactor plays a significant role in suppressing harmonic component and limiting the rising rate of short-circuit current [1]. With the rapid growth of HVDC power grid, the demand for dry-type smoothing reactor increases rapidly [2–4].

If lightning or short-circuit fault occur in Direct Current (DC) System, the reactor may withstand overvoltage. Due to the existence of stray capacitance, the voltage distribution is not uniform among reactors [5]. Serious overvoltage occurs at the first reactor, which may lead to an insulation failure and even affect the normal operation of DC system. Therefore, it is very important to determine the insulation test level of a single reactor and absolutely necessary to extract the stray capacitance of the reactor.

Literatures [6, 7] present the lightning overvoltage distribution among a plurality of smoothing reactors and stray capacitance of the reactor is needed. However, the article does not explain where the stray capacitance comes from. In order to extract accurate stray capacitance parameters, it is necessary to establish a model according to the actual structure of the smoothing reactor. Because the reactor consist of hundreds of conductors, the structure is complex. There is partial capacitance between every two conductors. The order of the capacitance matrix is so high that it is impossible to establish such a capacitance network in the software. To simplify the operation and analysis in transient analysis software, it is essential to simplify the capacitance matrix.

This paper focuses on the extraction and simplification of stray capacitance matrix of the UHVDC dry-type smoothing reactor, which is proved by simulation and measurement. The lightning overvoltage test voltage of a single reactor could be obtained through related simulations, which would provide a reference for the lightning overvoltage test of UHVDC dry-type reactor.

2 Simulation and Analysis

2.1 Theoretical Analysis

An independent system comprised of n + 1 conductors and the conductors are numbered from 0 to n in sequence. The quantity of corresponding charge are $q_0, q_1, \ldots, q_n$ and The charge quantity meets the following formula.

$$q_0 + q_1 + \ldots + q_k + \ldots + q_n = 0 \tag{1}$$

Conductor 0 is selected as zero potential point ($\varphi_0 = 0$). In a multi-conductor system with linear space medium, the relationship between the potential and the charge of each conductor is as follows:

$$\{\varphi\} = [\alpha]\{q\} \tag{2}$$

where $[\alpha]$ refers to potential coefficient matrix, which is only related to geometry, size, location and permittivity. The conductor charge is normally not given, but the potential of each conductor or the voltage between two conductors.

$$\{q\} = [\alpha]^{-1}\{\varphi\}=[\beta]\{\varphi\} \tag{3}$$

where β refers to inductance coefficient, which is only related to geometry, size, location and permittivity. β_{ii} and β_{ij} respectively stand for self-inductance coefficient and mutual inductance coefficient. β can be calculated by

$$\beta_{ij}=q_i\big/\varphi_j\big|_{\varphi_j\neq 0.} \text{ The other conductors are grounded and the potential is zero.} \tag{4}$$

In engineering analysis, the relationship between charge and potential is usually expressed by mutual capacitance between two conductors. For an independent system comprised of $n + 1$ conductors, it can be expressed as follows:

$$\left.\begin{aligned}
q_1 &= C_{1,0}(\varphi_1 - \varphi_0)+\ldots+C_{1,k}(\varphi_1 - \varphi_k)+\ldots C_{1,n}(\varphi_1 - \varphi_n) \\
&\cdots\cdots\cdots \\
q_k &= C_{k,1}(\varphi_k - \varphi_1)+\ldots+C_{k,0}(\varphi_k - \varphi_0)+\ldots C_{k,n}(\varphi_k - \varphi_n) \\
&\cdots\cdots\cdots \\
q_n &= C_{n,1}(\varphi_n - \varphi_1)+\ldots+C_{n,k}(\varphi_n - \varphi_k)+\ldots C_{n,0}(\varphi_n - \varphi_0)
\end{aligned}\right\} \tag{5}$$

where C_{i0} presents the self-partial capacitance; C_{ij} stands for the mutual partial capacitance. They can be calculated by

$$\left.\begin{aligned}
C_{i,0} &= \sum_{j=1}^{n} \beta_{ij} \\
C_{i,j} &= -\beta_{ij}
\end{aligned}\right\} i,j = 1, 2\ldots n \tag{6}$$

The working system of the smoothing reactor can be regarded as a static independent system. Its stray capacitance is related to the relative position, shape, size as well as space medium and has nothing to do with the operation state. If the reactor is fixed on the platform, the stray capacitance is constant.

2.2 Structure of a Small Reactor

The research object is a small hollow smoothing reactor, which consists of five coaxial cylinders. Each cylinder is twined by a wire and called as package. Each package is encapsulated by epoxy resin and glassfiber. The width of air flue between the packages is 27 mm. The Schematic view of the cross section of the reactor is displayed in Fig. 1.

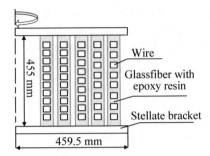

Fig. 1. Schematic view of the cross section of the reactor.

2.3   Extraction and Simplification of the Matrix

The number of turn of each package which belongs to the small smoothing reactor is different. One turn of wire is as a conductor. The first turn of wire of all packages is equipotential. The last turn of wire of all packages is also equipotential and is regarded as zero potential point. Each conductor is as a node and there are 42 nodes in total. There is a mutual capacitance between every two nodes and each node has a capacitance to the ground. The order of the reactor capacitance matrix is 41. The two-dimensional analysis model of the smoothing reactor is established through the finite element analysis software. The dielectric constant of epoxy glassfiber layer is 4.76. The resulting capacitance network matrix is displayed in Fig. 2 (non-adjacent mutual capacitance is not indicated) (Fig. 5).

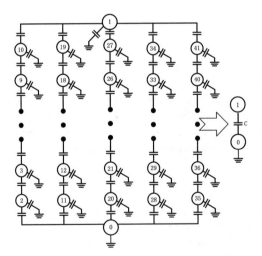

Fig. 2. Capacitance network of the reactor.

On the premise that the voltage between the two ends is kept unchanged, the end-to-end stray capacitance is obtained by simplifying the matrix. The relationship between potential and capacitance of each node is expressed as follows:

$$\left.\begin{aligned}
q_1 &= C_{1,0}(\varphi_1 - \varphi_0) + \ldots + C_{1,k}(\varphi_1 - \varphi_k) + \ldots C_{1,41}(\varphi_1 - \varphi_{41}) \\
&\ldots\ldots\ldots \\
q_k &= C_{k,1}(\varphi_k - \varphi_1) + \ldots + C_{k,0}(\varphi_k - \varphi_0) + \ldots C_{k,41}(\varphi_k - \varphi_{41}) \\
&\ldots\ldots\ldots \\
q_{41} &= C_{41,1}(\varphi_{41} - \varphi_1) + \ldots + C_{41,k}(\varphi_{41} - \varphi_k) + \ldots C_{41,0}(\varphi_{41} - \varphi_0)
\end{aligned}\right\} \quad (7)$$

where φ_0 refers to the potential of earth, $\varphi_0 = 0$; φ_1 is the given potential, $\varphi_1 = U$; $q_2 = 0, \ldots, q_{41} = 0$. The stray capacitance matrix in partial is displayed in Table 1.

The stray capacitance matrix is equivalent to an end-to-end stray capacitance C, which is calculated by

$$C = \frac{q_1}{\varphi_1 - \varphi_0} = 2.3223\,nF \quad (8)$$

The voltage at both the ends of the smoothing reactor is very low in reality. Therefore, the whole smoothing reactor is regarded as a conductor and considered to have one potential to the ground. In the example, the smoothing reactor is located on a platform which is 1.2 m away from the ground. The stray capacitance to the ground is 41.34693 pF, which is calculated by the finite element analysis software.

Table 1. Stray capacitance matrix in partial (Capacitance/pF).

Number	1	2	...	40	41
1	19.615	0.1814	...	9.2845	5424.51
2	0.1814	2285.68	...	0.0027	0.0028
...	...	...	...	...	...
40	9.2845	0.0027	...	2.32	5293.02
41	5424.51	0.0028	...	5293.02	2.317

2.4   Simulation Verification

The lightning overvoltage analysis model of three reactors in series is established in the transient analysis software and the simulation in accordance with the overvoltage distribution is carried out. Double exponential lightning waveform is adopted for simulation, which is 1.2/50 us and has an amplitude of 2100 kV [8–10]. Ea is the power waveform. E1, E2 and E3 are the voltage of the three series resistors, respectively. The two models are shown in Fig. 3. Figure 3a is the model of the capacitance matrix and Fig. 3b is the one of the equivalent stray capacitance. The waveform results are displayed in Fig. 4. The simulation results of the two models are listed in Table 2.

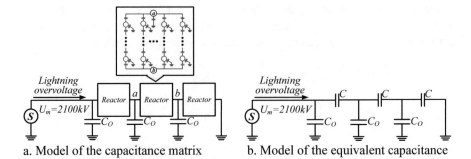

a. Model of the capacitance matrix   b. Model of the equivalent capacitance

Fig. 3. Transient analysis model of three reactors.

Table 2. Simulation results of the two models.

Number	1	2	3
1st model (kV)	717.65	693.18	681.05
2nd model (kV)	717.65	689.18	681.05

By comparing the simulation results of the two models, the results show that the lightning overvoltage distribution is almost the same, which indirectly proves that extraction and simplification method of the capacitance matrix is correct.

3   Experiment and Discussion

3.1   Extraction and Simplification of the Matrix of the Prototype

The dry-type smoothing reactor has a similar structure with the example, which is independently designed by our institute. It has 26 packages and consists of more conductors than the example. Each turn consists of 2 strands, 3 strands or 4 strands wires. The strands in the same turn are regarded as equipotential conductors. The schematic view of the cross section of the reactor is shown in Fig. 4.

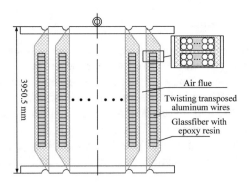

Fig. 4. Schematic view of the cross section of the reactor.

There is 5416 conductors in the reactor and the order of the capacitance network matrix is 5415. The two-dimensional analysis model of the smoothing reactor is established through the finite element analysis software. The dielectric constant of epoxy glassfiber layer is 4.76.

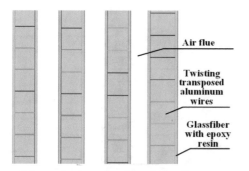

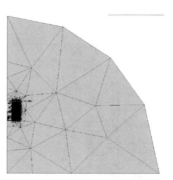

a. Schematic view of package in details   b. Schematic view of meshing of the air

Fig. 5. Schematic view of model and meshing of the reactor.

According to the capacitance matrix simplification method introduced above, the stray capacitance matrix is equivalent to an end-to-end stray capacitance C, which is calculated by

$$C = \frac{q_1}{\varphi_1 - \varphi_0} = 3.9201 \, nF \tag{9}$$

The smoothing reactor is located on a platform which is about 12 m away from the ground. The stray capacitance to the ground is calculated to be 1.536547 nF.

3.2   Experiment

The high frequency stray capacitance of the prototype was measured directly by an impedance analyzer [11]. The test leads should be as short as possible. The frequency ranged from 450 kHz to 500 kHz and the results are given in Table 3.

Table 3. Experiment results of the stray capacitance.

Frequency (Hz)	End-to-end stray capacitance (F)	Frequency (Hz)	End-to-ground stray capacitance (F)
453862	3.08E–09	453862	1.75E–09
455858	3.1E–09	455858	1.76E–09
457863	3.08E–09	457863	1.75E–09
459877	3.16E–09	459877	1.78E–09
461899	3.2E–09	461899	1.77E–09
463931	3.14E–09	463931	1.79E–09

(*continued*)

Table 3. (*continued*)

Frequency (Hz)	End-to-end stray capacitance (F)	Frequency (Hz)	End-to-ground stray capacitance (F)
465971	3.08E–09	465971	1.80E–09
468021	3.32E–09	468021	1.79E–09
470079	3.29E–09	470079	1.82E–09
472147	3.12E–09	472147	1.83E–09
474224	3.12E–09	474224	1.75E–09
476309	3.14E–09	476309	1.85E–09
478404	3.04E–09	478404	1.85E–09
480509	3.19E–09	480509	1.87E–09
482622	3.11E–09	482622	1.88E–09
484745	3.13E–09	484745	1.89E–09
486877	3.11E–09	486877	1.93E–09
489018	3.05E–09	489018	1.97E–09
491169	2.75E–09	491169	1.93E–09
493329	3.1E–09	493329	1.96E–09
495499	2.77E–09	495499	1.95E–09
497679	2.67E–09	497679	2.00E–09
499868	2.49E–09	499868	2.01E–09
Average (F)	3.05E–09	Average (F)	1.85E–09

Table 4. Simulation and experimental results.

	End-to-end stray capacitance (nF)	End-to-ground stray capacitance (nF)
Simulation	3.9201	1.536547
Experiment	3.05391	1.85
Error percentage (%)	28.5%	16.9%

Because the impedance analyzer is greatly disturbed by many kinds of external interference in the experiment, the error in measurement results is inevitable. The difference between simulation results and experimental results is acceptable. It shows that the method of extraction and simplification of the stray capacitance matrix is correct (Table 4).

4   Conclusion

This paper studies the method of the stray capacitance extraction and simplification of the dry-type smoothing reactor, which is proved to be correct by simulation and experiment.

The method introduced in this paper provides an important theoretical basis for obtaining stray capacitance of the same type of products. It is of great reference significance to verify interturn insulation of the reactor and determine the level of lightning overvoltage test of a single reactor as well as the test scheme.

References

1. Luo, Y.: Key technology analysis of dry-type smoothing reactors for ±800 kV UHVDC project. Water Resour. Power **30**(7), 159–162 (2012)
2. Yue, B., Zhang, Y., Zheng, J., et al.: Study on the technical specification of ±800 kV UHVDC dry-type smoothing reactor. High Voltage Eng. **32**(12), 170–173 (2006)
3. Zhang, J.: Selection of ultra-high voltage direct current smoothing reactors. Guangdong Electric Power **19**(5), 36–38 (2006)
4. Lei, Y., Zhao, X.: Parameter selection and arrangement of smoothing reactors for ±800 kV Yunnan-Guangzhou UHVDC transmission project. High Voltage Apparatus **48**(6), 75–79, 84 (2012)
5. Feng, W., Wei, X.: Analysis of characteristics of dry-type smoothing reactor in Yunnan-Guangdong ±800 kV UHVDC. Water Resour. Power **30**(10), 173, 143–145, 216 (2012)
6. Wu, Y., Guo, X., Yao, S., et al.: Study on distribution of impulse voltage between two UHVDC smoothing reactors. Electr. Equip. **8**(3), 20–22 (2007)
7. Yu, J., Lan, T., Chen, T., et al.: Overvoltage simulation analysis of lightning intrusion wave under different arrangement of smoothing reactor. Water Resour. Power **33**(6), 173, 195–198 (2015)
8. Shan, H., Peng, X., Wu, J., et al.: Analysis on lightning impulse type test for dry-type smoothing reactor of ±800 kV DC transmission demonstration project from Xiangjiaba to Shanghai. Power Syst. Technol. **34**(5), 12–15 (2010)
9. Chen, S., Yang, P., Geng, Y., et al.: Lightning overvoltage withstand of a ±800 kV smoothing reactor. J. Tsinghua Univ. (Sci. Technol.) **50**(1), 23–26 (2010)
10. Wang, Q., Li, W., Wang, B.: R & D of dry type smoothing reactor for ±800 kV HVDC power transmission. Electr. Equip. **7**(12), 11–14 (2006)
11. State General Administration of the People's Republic of China for Quality Supervision and Inspection and Quarantine, Standardization Administration of the People's Republic of China. GB/T 25092-2010 Dry-type air-core smoothing reactors for HVDC applications. China Standard Press, Beijing (2010)

Coordinated, Efficient and Optimized Crowd Evacuation in Urban Complexes

Huabin Feng[✉], Shiyang Qiu, Peng Xu, Shuaifei Song, Wei Zheng, and Hengchang Liu

University of Science and Technology of China, Suzhou, China
{fenghb,qsy1314,xp940626,songsf,weizheng}@mail.ustc.edu.cn,
hcliu@ustc.edu.cn

Abstract. With the development of the city, more and more urban complexes which are areas combine leisure, repast, and trade functions have emerged in our daily life. However, as the increasing of people in urban complexes, the danger of security incidents increases. Therefore, how to evacuate the crowd after a security incident as soon as possible has become a crucial issue. Evacuation not only refers to the process of transferring crowd from indoors to outdoors but also transfers the outdoor crowd to safe places by means of transportation, such as taxis, buses, subways and so on. To this end, we establish a human-vehicle collaborative evacuation system(HVCES) in Suzhou Center which is one of the largest and most advanced urban complexes in China. Based on the system, we obtain a safer and optimal vehicles scheduling strategy with heuristic algorithms to reduce the probability of an accident during the evacuation.

Keywords: Evacuation system · Vehicles scheduling · Heuristic algorithm

1 Introduction

With the rapid urbanization process of China, more and more urban complexes appear in cities. Since the urban complex has functions such as repast, shopping, and entertainment, a large number of citizens choose to spend their holidays in it. For example, Suzhou Center is the biggest urban complex in Suzhou, the total number of visitors can reach 500,000 on a busy day. In the case of so many people gathering, once a safety incident occurs, such as fire, explosion, earthquake, etc., it would lead to lots of casualties. Therefore, we need to transfer the people in the urban complex quickly through efficient transportation resource allocation after the accident so as to minimize the probability of accidents. There are various sensors in Suzhou Center, e.g., surveillance cameras are used for estimating crowd size and density, infrared sensors are used for counting the number of people entering and leaving, with iBeacon technology, we can get indoor positions of customers and navigate for them. With these data, we can not only provide customers with better service in ordinary but also provide a basis for rational resource scheduling to speed up evacuation under emergency.

© Springer Nature Switzerland AG 2020
J. H. Kim et al. (Eds.): ICHSA 2019, AISC 1063, pp. 148–159, 2020.
https://doi.org/10.1007/978-3-030-31967-0_17

There has been some work related to crowd evacuation. Helbing and Molnar proposed the social force model (SF model) [1] to study pedestrian movement behavior in crowded situations. In addition to the SF model, cellular automata model [2], lattice gas model [3], angent-based model [4], and reciprocal velocity obstacle model [5] are alse common in pedestrian simulation. All the models mentioned above belong to the micro model which can simulate pedestrian behavior well, but need calculating a route for each pedestrian is obligatory, so the computation amount of the entire simulation process is very huge. For each individual, the calculations include goal selection, plan computation, and plan adaptation. Goal selection means deciding which exit an individual wants to achieve, e.g., pedestrians always like to choose the nearest or familiar exit to escape [6]. Plan computation involves calculating a static path to arrive the target exit. Most path planning algorithms are graph-based algorithms based on the spatial discretization, e.g., delaunay triangulation [7], road maps [8], corridor maps [9], and navigation meshes [10]. These algorithms can not only respect to static obstacles but also work in dynamic change conditions. But generally speaking, it is assumed that there are no moving objects or pedestrians in the traversable space in the plan computation stage. Therefore, an individual needs to dynamically change the movement state according to the surrounding environment during the simulation process with plan adaptation. There are many models for plan adaptation, besides the models [1–5] we mentioned above, there are other approaches include vision based [11], continuum based [12], and rule-based [13]. In this paper, we choose the SF model for plan adaptation, and [14–16] studies the SF model on crowd evacuation in specific scenarios. However, these papers only work on how to speed up the crowd evacuation or studying the evacuation process itself, haven't involved the dispatch of vehicles to transport people, which will lead to a mass of people squeeze in the doorway. In essence, the danger is not fundamentally eliminated. Therefore, we add vehicles deployment strategies based on crowd evacuation to keep pedestrians away from dangerous areas.

The main contributions in our paper include: (a) completing the HVCES, including simulation of crowd indoors, route planning algorithm of pedestrians in emergency situations, and building the vehicles scheduling model; (b) presenting the evaluation criteria for the entire evacuation process and proposing the efficient and secure vehicles scheduling algorithm based on it; (c) verifying that compare with static vehicles scheduling, dynamically adjusting the vehicle schedule by using pedestrian evacuation information, the *alpha*-coefficient that indicates the degree of danger is reduced to 52.1%.

The remainder of this paper is organized in the following structure: in Sect. 2 we introduce the framework construction of HVCES system; in Sect. 3 we test the system and try various vehicles deployment strategies to get an optimal deployment strategy; and we close with our conclusion in Sect. 4.

2   System Framework

In order to simulate the pedestrian evacuation and vehicles scheduling in emergency situation, we established the HVCES. The HVCES is mainly composed of two parts: the pedestrian simulation model and the vehicles scheduling algorithm. With the pedestrian simulation, we obtain data on the number of people at the exit over time, and the vehicles scheduling algorithm transfers them to other places. The entire system framework is shown as Fig. 1.

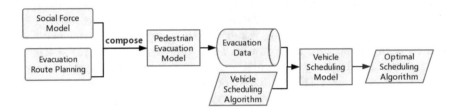

Fig. 1. System framework

2.1   Pedestrian Evacuation Model

Social Force Model. SF model is a commonly used model in crowd evacuation, it divides the power of human movement into three parts: the desired force for the targe $\boldsymbol{f}_{des}$, the repulsive force between people $\boldsymbol{f}_{soc}$, and the resistance of the obstacle $\boldsymbol{f}_{obs}$. Assume there is a pedestrian denoted as P, the weight of P is m. At time t, his speed is $\boldsymbol{v}(t)$, and τ is a quiet small time interval, so the SF model can be expressed as Eq. 1.

$$m\frac{\boldsymbol{v}(t) - \boldsymbol{v}(t-\tau)}{\tau} = \boldsymbol{f}_{des} + \boldsymbol{f}_{soc} + \boldsymbol{f}_{obs} \tag{1}$$

Suppose the current position of P is point P1, his target position is point P2, and $\boldsymbol{e}(t)$ represents the unit vector from P1 to P2. The current velocity of him is $\boldsymbol{v}(t)$, and his maximum walking speed is v_m, the mathematical expression of $\boldsymbol{f}_{des}$ is shown as Eq. 2.

$$\boldsymbol{f}_{des} = m\frac{v_m\boldsymbol{e}(t) - \boldsymbol{v}(t)}{\tau} \tag{2}$$

The SF model can describe the psychological tendency of two strangers that they always want to stay away from each other. Suppose in the vicinity of P there are n people who will make this effect on him, and they form a set denoted as S_n. Besides, we define four constants—A_i, B_i, k, and κ. A_i indicates the strength of the force, B_i indicates the effective distance of force. $\boldsymbol{n}_j$ represents a unit vector from the position of the jth people in S_n to P's position. $\boldsymbol{t}_j$ is the orthogonal unit vector of $\boldsymbol{n}_j$. In our system, everybody is considered as a circle, the radius of P is r_i, the radius of jth people in S_n is r_j, and $r_{ij} = r_i + r_j$. In addition, d_{ij} indicates the distance between the centers of them, $\boldsymbol{v}_i$ is P's velocity, $\boldsymbol{v}_j$ represent the velocity of jth people in S_n, and $\Delta\boldsymbol{v}_{ji} = \boldsymbol{v}_j - \boldsymbol{v}_i$.

At last, we use $g(x)$ to indicate ramp function whose formula is $g(x) = \max(0, x)$. So the equation of $\boldsymbol{f}_{soc}$ is:

$$\boldsymbol{f}_j = \sum_{j \in S_n} \{[A_i \exp(\frac{r_{ij} - d_{ij}}{B_i}) + kg(r_{ij} - d_{ij})]\boldsymbol{n}_j + \kappa g(r_{ij} - d_{ij})(\varDelta\boldsymbol{v}_{ji} \cdot \boldsymbol{t}_j)\boldsymbol{t}_j\} \quad (3)$$

$\boldsymbol{f}_{obs}$ is the effect of obstacles on P motion, but not all obstacles can affect P, so m obstacles who can affect P form a set denoted W_m. Suppose w is an obstacle in W_m, d_{iw} indicates the distance from the center of P to w, $\boldsymbol{n}_{iw}$ indicates the unit vector perpendicular to the surface of w pointing to P, and $\boldsymbol{t}_{iw}$ is the orthogonal unit vector of $\boldsymbol{n}_{iw}$. So the equation of $\boldsymbol{f}_{obs}$ is:

$$\boldsymbol{f}_{obs} = \sum_{w \in W_m} \{[A_i \exp(\frac{r_i - d_{iw}}{B_i}) + kg(r_i - d_{iw})]\boldsymbol{n}_{iw} - \kappa g(r_i - d_{iw})(\boldsymbol{v}_i \cdot \boldsymbol{t}_{iw})\boldsymbol{t}_{iw}\} \quad (4)$$

Evacuation Route Planning. In addition to pedestrian behavior simulation, we also need to plan for escape routes. In the case of emergency, a person always chooses the nearest exit to escape, so our route planning algorithm should reflect this behavior.

Suppose there are k exists represented by the set S_k. We grid the simulation scene into n rows × m columns, and store the grid information into an array named *Info*. Each element in Info has two attributes — D_e and *nextHop*, D_e indicates the distance from the corresponding grid point to an exit whose initial value is infinity, and *nextHop* indicates the next grid point in the route to the exit. We use the breadth-first search(BFS) to update *Info*, the detail process is shown in Algorithm 1.

Algorithm 1. Route planning with BFS

1: Initialize S_k and $Info$
2: **for** $i = 1 \rightarrow k$ **do**
3:     $P \leftarrow S_k[i]$
4:     initialize an empty queue Q
5:     push P to the end of Q
6:     **while** Q isn't empty **do**
7:         $tp \leftarrow$ pop front element of Q
8:         $neighbors \leftarrow$ getNeighborGrids(tp)
9:         **for all** q in $neighbors$ **do**
10:             $D_{qp} \leftarrow$ distance from q through tp to P
11:             **if** $D_{qp} < Info[q].D_e$ **then**
12:                 $Info[q].D_e \leftarrow D_{qp}$
13:                 $Info[q].nextHop \leftarrow q$
14:                 push q to the end of Q

After getting $Info$, for any valid position in the scene, we can recursively get the next hop to form an escape route.

2.2   Vehicles Scheduling Model

There are 16 bus lines near the Suzhou Center. During the rush hour, the bus will be dispatched each 5–8 min, that is one bus passing in every 20 s–30 s. If we use the buses as evacuation vehicles, with the scheduling of command center it can be guaranteed that there is at least one bus arrives in every 10 s.

A bus can carry up to 80 people. Since there may be several people on it, we assume that the bus can still carry 50 people. By observing at the bus station, we find if the front door and back door of a bus are opened at the same time, boarding speed can be up to 3 people per second.

Therefore, our vehicles scheduling model is: every 10 s, there is a bus that can carry 50 people to one exit of the Suzhou Center. The specific exit is determined by the scheduling algorithm. After bus arriving, the pedestrians get on the bus at a rate of 3 people per second. If there are n buses at the same exit, the total boarding speed is $3n$ people per second, and a bus would leave after loading 50 people.

2.3   Evaluation Index

In order to evaluate the scheduling strategy, we define α coefficient to quantify the degree of danger at the doorways. It is well known that the more people there are in a fixed area place, the more dangerous it is, and when there are few people, the probability of an accident is almost zero. So the mathematical expression of α coefficient is shown as Eq. 5.

$$\alpha(n) = k_1 \exp\left(\frac{n - N_s}{N_s} k_2\right) \tag{5}$$

The formula above represents the relationship between the people number n and α coefficient at time t at the specific doorway. The N_s represents the safety threshold of the people count. When n is less then N_s, $\alpha(n)$ is close to 0. After n greater than N_s, $\alpha(n)$ increases rapidly with the increase of n. If the area of the doorway is larger, the value of N_s is larger. k_1 and k_2 are constants used to adjust the α coefficient curve. In our experiment, we set $N_s = 100$, $k_1 = 0.01$ and $k_2 = 3$. The α coefficient curve is shown as Fig. 2, the red line indicates the value of N_s.

If we want to assess the degree of danger at the ith doorway in the evacuation process that begins from time $t0$ and ends to time $t1$, we only need to calculate the integral of α coefficient over time that is shown as Eq. 6. In the equation, $n_i(t)$ indicates people count of the ith doorway at time t.

$$D_i = \int_{t0}^{t1} \alpha\big(n_i(t)\big)\, dt \tag{6}$$

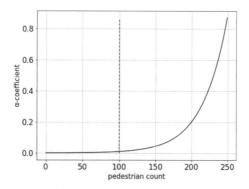

Fig. 2. Alpha coefficient curve

3 Evaluation

In the evaluation, we apply the evacuation route planning algorithm to the Suzhou Center and conducted an indoor crowd evacuation simulation, then we come up 3 vehicles scheduling algorithms and test them based on the evacuation data. In the end, we get the optimal scheduling algorithm and validate it in various scenarios.

3.1 Evacuation Route Planning

We apply the route planning algorithm mentioned in Sect. 2.1 to the Suzhou Center, the exits in it are shown in Fig. 3. And Fig. 4 is the exits selection distribution map, the same color in it means the same selection.

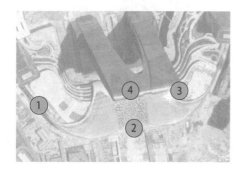

Fig. 3. Four exits in Suzhou center

Fig. 4. Exits selection map

3.2   Indoor Crowd Evacuation

The parameters of SF model mentioned in Sect. 2.1 are set as follows: for a pedestrian, $m = 60\,\mathrm{kg}$, $r_i = 0.3\,\mathrm{m}$, $\tau = 0.1\,\mathrm{s}$, $v_m = 1.5\,\mathrm{m/s}$, $A_i = 200$, $B = 0.1$, $k = 1.2 \times 10^5$, $\kappa = 2.4 \times 10^5$, and treat distance less than $3\,\mathrm{m}$ as neighbors.

We arrange 1000 people in Suzhou Center, the distribution of pedestrians is shown in Fig. 5.

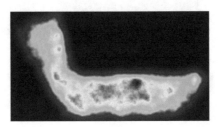

Fig. 5. Pedestrian density map

With the simulation of SF Model, we obtain the curve of the people counts at the four exits shown as Figs. 6, and 7 is the pie chart of them. From Fig. 7 we can see that the ratio of people going out from the four exits is about exit-1:exit-2:exit-3:exit-4=2:2:1:2.

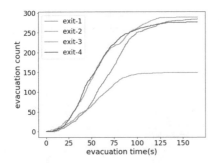

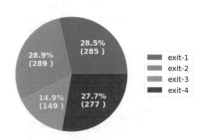

Fig. 6. People number without bus dispatch

Fig. 7. People evacuation ratio

3.3   Vehicles Scheduling Algorithm

We apply 3 vehicles scheduling strategies to the above-mentioned crowd evacuation model and evaluate with α coefficient.

Round-Robin Scheduling Algorithm. If we can't get the real-time number of people at each doorway, we can only assign vehicles with the round-robin scheduling(RRS) algorithm. Assuming the rotation order is exit-1→exit-2→exit-3→ exit-4, curves of the people at exits shown as Fig. 8.

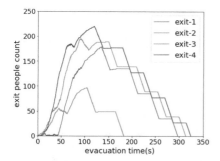
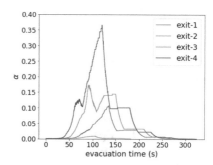

Fig. 8. People number with RRS   **Fig. 9.** α Coefficient with RRS

According to Eq. 5, we get the α coefficient curves shown in Fig. 9. As can be seen from Fig. 9, the α coefficient curve of exit-4 will reach a peak of about 0.37. With Eq. 6 we calculate the α coefficient throughout the evacuation process, they are 6.136, 13.617, 0.531, 23.007, and their average is 10.823. It is obvious that α coefficient of exit-4 is much larger than others. Therefore, the round-robin scheduling algorithm isn't a good deployment strategy.

People-Number First Algorithm. The algorithm of Sect. 3.3 hasn't used the information of people count at the doorways. However, high-definition cameras are installed at each doorway of Suzhou Center, and the number of people can be counted by computer vision technology. So we can use this extra information for better scheduling.

The main idea of people-number first(PNF) algorithm is to dispatch the vehicles to the doorway whose number of people is most, very similar to the idea of greedy algorithm. With this algorithm, the curves of the people number at outdoors is shown as Fig. 10. And Fig. 11 is the corresponding α coefficient curves in Fig. 10.

Comparing Fig. 9 with Fig. 11, we can see that the latter α coefficient of exit-4 is significantly smaller than the former. And the results of Eq. 6 are 8.327, 4.769, 4.163, 5.301, and their average is 5.640 that drops to 52.1% of the RRS algorithm.

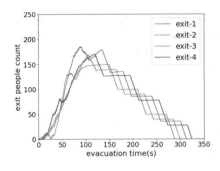
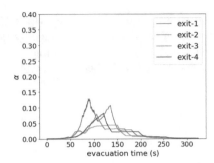

Fig. 10. People number with PNF   **Fig. 11.** α Coefficient with PNF

Beam Search Algorithm. PNF algorithm significantly reduces the overall α coefficient compared to RRS algorithm. However, we need to determine how much promotion space PNF algorithm has relative to the best scheduling algorithm.

From Fig. 6, we can calculate the number of rescue bus required for each exit that the exit-1, exit-2, and exit-4 each required 6 buses and exit-3 required 3 buses. Therefore, the scheduling strategy can be regarded as an arrangement of 6 ones, 6 twos, 6 fours, and 3 threes, just as a gene fragment in the genetic algorithm(GA). We use the gene mutation method in GA to get new scheduling strategies and obtain the global optimal scheduling strategy by beam search(BS) algorithm which is a heuristic search algorithm. The flow chart of it is shown in Fig. 12.

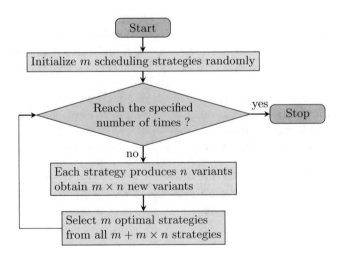

Fig. 12. Beam search optimal scheduling strategy

We set m = 10, n = 5, and the specified number of times as 50. With BS algorithm, we determine that the minimum value of the 4 exits α mean is 5.548. The result of the PNF algorithm is 5.640 which is very close to the minimum value, so the PNF algorithm is acceptable.

3.4 Robustness Verification

In order to verify the robustness of the PNF algorithm, we conduct experiments on different people distributions. The pedestrian density maps of the two experiments are shown in Figs. 13 and 14.

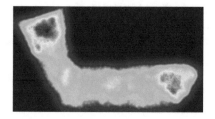
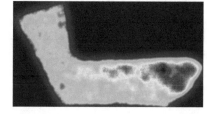

Fig. 13. Density map of experiment 2   **Fig. 14.** Density map of experiment 3

The results of the three experiments including the experiment in Sect. 3 are shown in Table 1. The first to fourth columns are the evacuation ratios of each exit, and the fifth behavior is the α coefficient mean value.

As can be seen from the table, although the distribution of pedestrians is variant, the values of α coefficient mean are all close. Therefore, our conclusion is that the PNF algorithm is effective for various distributions of pedestrian.

Table 1. Results of the three experiments

	Exit-1(%)	Exit-2(%)	Exit-3(%)	Exit-4(%)	α-mean
1	28.5	28.9	14.9	27.7	5.640
2	38.1	21.8	20.5	19.6	5.903
3	25.1	25.3	24.9	24.7	6.119

3.5 Discussion

In this section, we conduct an indoor evacuation simulation at the Suzhou Center and obtain crowd evacuation data, and we apply RRS algorithm, PNF algorithm, and BS algorithm to vehicles scheduling. The α-coefficient of PNF algorithm is about only half of RSS algorithm. The global optimal scheduling strategy obtained by the BS algorithm is about little better than the PNF algorithm, but requires a lot of calculation and data of the entire evacuation process which results in its scheduling strategy not being generated in real time. So, the PNF algorithm is the optimal algorithm trades off effect and real-time.

4   Conclusion

In this paper, we combine the indoor crowd evacuation model with the outdoor vehicles scheduling algorithms to study how to safely and efficiently evacuate people in urban complex area. We propose α coefficient to evaluate the risk level of the evacuation process, and we prove that the PNF algorithm is significantly better than the RRS algorithm by experiments. In addition, we prove that the effect of PNF algorithm is very close to the optimal scheduling by mean of beam search. Finally, we show that the PNF algorithm is effective for different pedestrian distributions.

References

1. Helbing, D., Molnar, P.: Social force model for pedestrian dynamics. Phys. Rev. E Stat. Phys. Plasmas Fluids Relat. Interdisc. Top. **51**(5), 4282 (1998). https://doi.org/10.1103/PhysRevE.51.4282
2. Zhou, X., Jingjie, H., Xiongziyan, X.: Cellular automaton simulation of pedestrian flow considering vision and multi-velocity. Physica A S0378437118311786 (2018). https://doi.org/10.1016/j.physa.2018.09.041
3. Tao, Y.Z., Dong, L.Y.: A floor field real-coded lattice gas model for crowd evacuation. Epl **119**(1), 10003 (2017). https://doi.org/10.1209/0295-5075/119/10003
4. Tan, L., Hu, M., Lin, H.: Agent-based simulation of building evacuation: combining human behavior with predictable spatial accessibility in a fire emergency. Inf. Sci. **295**, 53–66 (2015). https://doi.org/10.1016/j.ins.2014.09.029
5. Bera, A., Manocha, D.: Realtime multilevel crowd tracking using reciprocal velocity obstacles (2014). https://doi.org/10.1109/ICPR.2014.714
6. Kinateder, M., Comunale, B., Warren, W.H.: Exit choice in an emergency evacuation scenario is influenced by exit familiarity and neighbor behavior. Saf. Sci. **106**, 170–175 (2018). https://doi.org/10.1016/j.ssci.2018.03.015
7. Qureshi, A.H., et al.: Re-planning using delaunay triangulation for real time motion planning in complex dynamic environments. In: 2018 IEEE/ASME International Conference on Advanced Intelligent Mechatronics (AIM). IEEE (2018)
8. Gonzalez, D., et al.: A review of motion planning techniques for automated vehicles. IEEE Trans. Intell. Transp. Syst. **17**(4), 1–11 (2015). https://doi.org/10.1109/TITS.2015.2498841
9. Geraerts, R., et al.: Using the corridor map method for path planning for a large number of characters. Motion in Games. Springer Berlin Heidelberg (2008). https://doi.org/10.1007/978-3-540-89220-5_2
10. Oliva, R., Pelechano, N.: Automatic Generation of Suboptimal NavMeshes. In: International Conference on Motion in Games Springer, Berlin (2011). https://doi.org/10.1007/978-3-642-25090-3_28

11. Ondrej, J., Pettr, J., Olivier, A.-H., et al.: A synthetic-vision based steering approach for crowd simulation. ACM Trans. Graphics **29**(4), 1 (2010). https://doi.org/10.1145/1833351.1778860
12. Narain, R., et al.: Aggregate Dynamics for Dense Crowd Simulation. ACM Trans. Graphics **28**(5), 1–8 (2009). https://doi.org/10.1145/1618452.1618468
13. Masuda, S., Nakamura, H., Kajitani, K.: Rule-based searching for collision test cases of autonomous vehicles simulation. IET Intell. Transp. Syst. **12**(9), 1088–1095 (2018). https://doi.org/10.1049/iet-its.2018.5335
14. Hou, L., et al.: A social force evacuation model with the leadership effect. Physica A 400(4), 93-99. https://doi.org/10.1016/j.physa.2013.12.049
15. Wan, J., Sui, J., Yu, H.: Research on evacuation in the subway station in China based on the Combined Social Force Model. Physica A **394**, 33–46 (2014). https://doi.org/10.1016/j.physa.2013.09.060
16. Luh, P.B., et al.: Modeling and optimization of building emergency evacuation considering blocking effects on crowd movement. IEEE Trans. Autom. Sci. Eng. **9**(4), 687–700 (2012). https://doi.org/10.1109/tase.2012.2200039

The Application of Big Data to Improve Pronunciation and Intonation Evaluation in Foreign Language Learning

Jin-Wei Dong, Yan-Jun Liao, Xiao-Dan Li, and Wei-bo Huang[✉]

Guangdong University of Foreign Studies,
Guangzhou 510006, Guangdong, People's Republic of China
172564946@qq.com

Abstract. The key to Foreign Language Learning lies in the accurate mastery of pronunciation and intonation. However, there exists a lack of learning environment and awareness of pronunciation practice in China. Most oral foreign language learning software can only provide fuzzy evaluation for learners. And learners cannot get detailed feedback to correct their pronunciation errors. Our objective in this study is to improve the evaluation function of pronunciation and intonation in the oral English learning system with the help of Big Data on education. To provide detailed feedback for learners, we set five evaluation indexes: pronunciation, intonation, fluency, sentence stress and rhythm. Our program used GOP (Goodness of Pronunciation), Duration Score and spectrogram from Praat to evaluate learner's oral English. The experimental results suggest that the evaluation function proposed in this paper is more effective than the traditional one. The accuracy of learners' pronunciation and intonation improves more effectively.

Keywords: Big data · Evaluation of pronunciation and intonation · Foreign language learning

1 Preface

Foreign Language Learning includes four aspects: listening, speaking, reading and writing. As an important part of language structure, pronunciation and intonation are the starting point of teaching and the key to learning English well. Gui Cankun (1985), a famous English phonetician in China, believed that "Pronunciation is the most important basic skill to learn English well" [1]. To a large extent, phonetic competence reflects the language competence of foreign language learners. In addition, pronunciation and intonation in communication often convey the moods and attitudes of the speakers. If pronunciation and intonation are used improperly, it will often cause misunderstandings. However, most Chinese lack timely and effective sources of feedback on the quality of pronunciation and intonation in oral English practice. When mistakes cannot be corrected in time, it will be not conducive to improving spoken English [2]. With the development of technology, there are more and more learning platforms. The advent of Big Data also brings opportunities for English pronunciation and intonation learning [3, 4].

© Springer Nature Switzerland AG 2020
J. H. Kim et al. (Eds.): ICHSA 2019, AISC 1063, pp. 160–168, 2020.
https://doi.org/10.1007/978-3-030-31967-0_18

2   Current Situations of Oral Foreign Language Learning Platforms

Some oral English learning applications such as "Laix", "Interesting English Dubbing" and "Youdao Oral Master" only gives a whole fuzzy score or points out the error of a single word. It is difficult for learners to find out their problems in pronunciation and intonation due to lack of detailed feedback. Therefore, learners' oral English level cannot be improved efficiently. In the big data era, education should keep up with the times. The 4Vs of big data Volume, Variety, Velocity and Veracity can help make decisions. Learning platforms need to make use of big data on education to collect learners' learning information to provide timely and personalized feedback. Big data on education means the data of learners' behavior obtained from learning platforms in narrow sense [3]. Excavating and utilizing learners' learning data to improve teaching efficiency and effectiveness has become the trend of phonetics teaching. There should be five indexes in the evaluation system: pronunciation, intonation, fluency, sentence stress and rhythm [5]. Accurate guidance can help learners to correct errors.

The construction of intelligent oral foreign language learning and evaluation system based on computer and network has become reality. Some scholars have developed some more mature systems. However, these research results used a variety of identification methods without a unified scheme. Most systems are available only for large-scale phonetics laboratories or English test systems. There are some social needs for portable mobile handset software for individual learners [6]. Thus, we need to improve the pronunciation and intonation evaluation function of oral foreign language learning systems. We try to provide detailed feedback for learners and help them correct errors in pronunciation and intonation.

3   Speech Recognition Technology and Principles of Evaluation

Automatic Speech Recognition (ASR) is the use of computer techniques to identify and process human voice. It is used to identify the words a person has spoken and convert speech signals into appropriate commands or texts. In recent years, the combination of Deep Neural Network (DNN) and the traditional Hidden Markov Model (HMM) have made great achievements in speech recognition. Spoken language evaluation technology based on DNN has attracted much attention of the Industry and become the main direction in the research of Computer-assisted Language Learning (CALL) [7].

Speech is segmented into phonemes according to given a text by using automatic speech recognition technology. Based on it, the scoring features which can better reflect the accuracy and fluency of pronunciation are evaluated. Then by running the detection algorithm with the predetermined model, the system can get the final scores of pronunciation and intonation. There are some common score indexes. Among them, the frame-normalized log posterior probability is the most effective feature for the evaluation on pronunciation. The rate of speech (ROS) and the duration scores reflect the fluency of pronunciation. The acoustic model is the most important basis for calculating the frame-normalized log posterior probability. Therefore, in the evaluation of speech, we usually use standard pronunciation to build acoustic models. And the Goodness of Pronunciation (GOP) is used to simplify the theory of frame-normalized log posterior probability [8].

The evaluation of intonation, sentence stress and rhythm can be realized through Praat. Praat is a scientific software program for the analysis of speech in phonetics. Users can also run Praat programs on other operating systems after modifying and compiling the source codes. Praat can record sound signals and extract parameters such as fundamental frequency, resonance peak, pitch, intensity, length, etc. [9] Praat can also draw a phonological diagram and a high-and-low arch of intonation. Through these sonograms, we can find out the characteristics of pronunciation in a speech segment, such as resonance peak (tongue position), clarity and vowel length. The pitch contour can show the track of intonation. Learners' intonation, sentence stress and rhythm can be easily compared with the standard one.

4 Overall Framework Design on the System

The system consists of three main parts, management part, application part and evaluation part. The overall framework of the oral English learning system assisted with big data techniques, as is shown in Fig. 1.

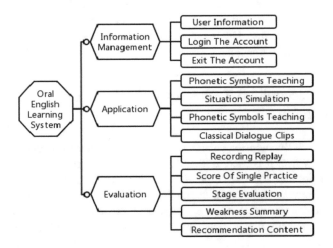

Fig. 1. Overall framework of oral english learning system.

The user information management part conserves all the account information registered by users. The users can modify their information, login or exit their accounts in this part.

There are four main functions in the application part. The first one function is phonetic symbols teaching. Learners can follow the teaching audios to learn the pronunciation of English phonetic symbols, which helps make pronunciation more standard.

The situation simulation function covers most of the situations in life and work. Learners can practice conversations in different situations. The abundant online corpora can strengthen the practice of everyday speech. There are also a lot of dialogue clips from classic films and television works. By simulating these clips, learners will find more interest in learning oral English.

The last part is the evaluation part in which the system scores and evaluates learners' pronunciation and intonation. Learners replay their recordings and compare them with the standard audios to find their shortcomings directly. Oral evaluation is divided into two main aspects. One is the scores to be displayed immediately after each single oral practice, which is convenient for learners to correct errors according to feedback in time. The other is the summary evaluation of each stage. This kind of evaluation is good for learners to draw lessons from the past and determine the learning direction for next stage. According to the stage evaluation, the system automatically summarizes learners' weaknesses and provides personalized guidance for learners. Learners can strengthen exercises based on their weaknesses. Meanwhile, the system automatically recommends the content of the exercises. Learners do not have to search for what they need in a vast amount of practice materials by themselves.

5   The Process of Evaluation on Pronunciation and Intonation

This paper gives a brief introduction to the overall framework of the oral English learning system. The key part is the concrete operation flow of the evaluation on pronunciation and intonation, as is shown in Fig. 2.

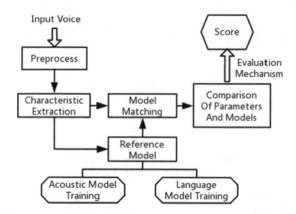

Fig. 2. The procession of pronunciation and intonation evaluation.

The first step is to pre-process and digitize the voice signals. Then the system makes the characteristic extraction of the input voice signals. The third step is to build the acoustic models and language models, and to modify the parameters of these two models with MSE (Mean Square Error Criterion). Finally, the score is calculated by matching parameters and models [10].

To provide detailed feedback for learners, we set five evaluation indexes: pronunciation, intonation, fluency, sentence stress and rhythm, as shown in Fig. 3.

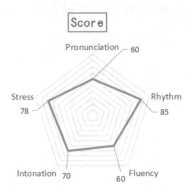

Fig. 3. The scores of five evaluation indexes.

The system calculates the scores based on these five indicators. Learners can know about their scores through the radar diagram and find out their weaknesses in pronunciation and intonation directly.

5.1 Evaluation on Pronunciation

Firstly, the system transforms the input from learners into its corresponding phonemic sequence. Then we get small segments of speech corresponding to each phoneme. Next, the quality of pronunciation of each small segments of speech can be evaluated by algorithms. In this paper, we quote the GOP (Goodness of Pronunciation) proposed by Witt and Young to evaluate.

$$
\begin{aligned}
\text{GOP(p)} &= \frac{logP(p|O)}{FrameCount(O)} \\
&= \frac{logP(p|O) - log\sum_{q \in Q} P(O|q)}{FrameCount(O)} \\
&\approx \frac{logP(O|q) - max(log\sum_{q \in Q} P(O|q))}{FrameCount(O)}
\end{aligned}
\tag{1}
$$

The "O" represents a given piece of speech. The "p" represents a phoneme in this speech fragment. The "P (O | p)" denotes the probability of "O" generated by the acoustic model of phoneme "p". The "P (O | q)" is the probability that a sound fragment belongs to a phoneme "p" generated by any phoneme "Q". The "Q" refers to the collection of all phonemes. To make the GOP of every phoneme relatively comparable, the system frame-normalize "O" as its frame-count. "Frame-count (O)" means the duration of pronunciation. Finally, the system judges whether the pronunciation is correct by comparing GOP with a predetermined threshold.

5.2   Evaluation on Fluency

Speech speed varies in different emotional states. Therefore, in the oral foreign language learning, learners usually need to imitate the speed of a standard example. The fluency of English pronunciation can be evaluated by the speed of speech. We compare the length of test audios with the standard reference audios. In this way we can get the scores about fluency. At first, the audios are detected with the double-gate threshold point detection method by using short-time energy and short-time average zero-crossing rate parameters. Phonetic syllables are divided according to this method. Then the length of effective pronunciation of a sentence (Len) is calculated by summing up the length of every syllable. The "Len (test)" refers to the duration of test audios. The "Len (std)" refers to the duration of standard audios. The ratio of duration (Duration Score) is the measurement of fluency evaluation.

$$\mathrm{DurationScore} = \frac{Len_{test}}{Len_{std}} \tag{2}$$

5.3   Evaluation on Intonation, Stress and Rhythm

Sonograms drawn by Praat can help learners to review nucleus placement, the distribution of stress and the intonation of a sentence. The system compares audio data input of learners with the standard audio data. According to these sonograms, the evaluation of intonation, stress and rhythm can be realized.

6   Experimental Results

To test the effectiveness of the evaluation function with the help of big data, we experimented and confirmed the system. Thirty non-English majors' undergraduates in Guangdong University of Foreign Studies were divided into two groups (group A and group B). Group A is the control group. Group B is the experimental group. Every group has 15 undergraduates. They all learn English as a second foreign language. The experiment was held in the same place. All the undergraduates were provided with the same recording equipment. They were required to practice pronunciation and intonation with the same learning materials at the same time. The oral English learning system used by group A is the traditional system. The traditional system can only provide a fuzzy score for learners. Meanwhile, group B was provided with the improved system. The improved system has more functions in evaluation on pronunciation and intonation.

In the process of the experiment, two groups practiced the same content. At the first predetermined time, two groups evaluated their pronunciation and intonation with the help of the provided system. After getting their scores and feedback, all the undergraduates practiced the same sentences and corrected their errors. Finally, their pronunciation and intonation were tested again. Then the differences between the scores of these two times are calculated. In this way, the effectiveness of errors correcting different evaluation methods can be compared.

All the content used in the experiment are the classical lines in movies or television clips:

(1) You are never wrong to do the right thing.
(2) Whatever you do, do it a hundred percent.
(3) Do not obsess over the way of life's problems.
(4) You will never know what you can do till you try.
(5) It does not do to dwell on dreams and forget to live.

Since the original English proficiency of every student was different, we choose the average improved rate as the indicator for comparing the effectiveness of two evaluation methods, as is shown in Figs. 4 and 5.

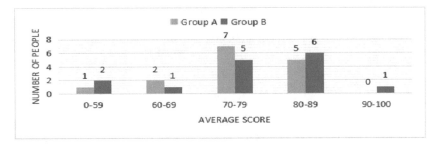

Fig. 4. The average score of the first evaluation.

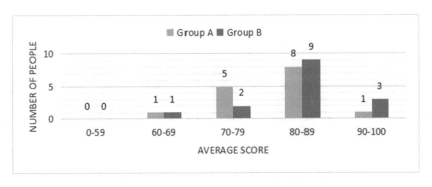

Fig. 5. The average score of the second evaluation.

The results of the 2 tests show that the average improved rate of the experimental group (group B) on improving the pronunciation and intonation is 13.12%. The average improved rate of the control group (group A) is 9.54%. The improved rate of Group B is obviously higher than that of group A. Through the comparison of the results of these two groups, we can see that the improved system is more effective. The evaluation function on pronunciation and intonation is improved with the help of big data. Through five evaluation indexes, learners get detailed scores and feedback. Learners can correct their errors effectively and improve the accuracy of pronunciation and intonation.

7 Summary

In the tide of globalization, English will never be short of learning and users. The key to learning English well is to master the pronunciation and intonation accurately. However, there is no widely used portable pronunciation and intonation evaluation system for learners. Learners cannot accurately understand their weaknesses because they lack detailed and timely feedback. Based on big data we improve the evaluation function of pronunciation and intonation in the spoken English learning system. The system uses big data technology to extract useful information. It combines the GOP, Praat and other tools to evaluate learners' oral English through five indexes. Learners can get personalized guidance. But this function is not yet complete. Technical improvements are needed for large-scale usage of this system. Therefore, in the era of big data, we should seize the opportunity to improve the evaluation function on the pronunciation and intonation continuously.

Acknowledgement. This study was financially supported by the Undergraduate Innovation Training Project of Guangdong University of Foreign Studies in 2019 (NO. 201911846007).

References

1. Chun, Z.Y.: Reflections on the Research of College English Phonetics Teaching Based on Big Data. Crazy English (Pro) 107–108 (2018)
2. Laleye, F.A.A., Ezin, E.C., Motamed, C.: Toward an automatic fongbe speech recognition system: hierarchical mixtures of algorithms for phoneme recognition. In: Madani, K., Peaucelle, D., Gusikhin, O. (eds) Informatics in Control, Automation and Robotics. Lecture Notes in Electrical Engineering, vol 430, pp. 133–149. Springer, Cham (2018)
3. Huang, W.B., Ruan, L.X., Liu, J.H., Li, X.D.: Adaptive learning system for foreign language writing based on big data. In: Hao T., Chen W., Xie H., Nadee W., Lau R. (eds) Emerging Technologies for Education. SETE 2018. Lecture Notes in Computer Science, vol 11284, pp. 12–22. Springer, Cham (2018)
4. Liu, J.H., Liang, W.X., Li, X.D.: Formative assessment system of college english based on the big data. In: Xiong, N., Xiao, Z., Tong, Z., Du, J., Wang, L., Li, M. (eds) Advances in Computational Science and Computing. ISCSC 2018 2018. Advances in Intelligent Systems and Computing, vol 877, pp. 472–480. Springer, Cham (2019)
5. Garcia, M.T., Bargh, J.A.: Automatic evaluation of novel words the role of superficial phonetics. J. Lang. Soc. Psychol. 22(4), 414–433 (2003)
6. Wang, S.: Design and implementation of intelligent english pronunciation training system based on android platform. Nanjing University of Posts and Telecommunications (2013)
7. Millor, N., Lecumberri, P., Gomez, M., Martinezramirez, A., Rodriguezmanas, L., Garciagarcia, F.J., et al.: Automatic evaluation of the 30-s chair stand test using inertial/magnetic-based technology in an older prefrial population. IEEE J. Biomed. Health Inform. 17(4), 820–827 (2013)

8. Yan, K.: Acoustic model optimization for automatic pronunciation quality assessment. In: International Conference on Multisensor Fusion & Information Integration for Intelligent Systems (2014)
9. Ruzanski, E., Hansen, J.H.L., Meyerhoff, J., Saviolakis, G.: Effects of phoneme characteristics on TEO feature-based automatic stress detection in speech, (ICASSP 2005). In: IEEE International Conference on Acoustics, Speech, and Signal Processing (2015)
10. Duan, R., Kawahara, T., Dantsujii, M., Zhang, J.: Pronunciation error detection using DNN articulatory model based on multi-lingual and multi-task learning. In: International Symposium on Chinese Spoken Language Processing (2017)

Serviceability Assessment Model Based on Pressure Driven Analysis for Water Distribution Networks Under Seismic Events

Do Guen Yoo$^{(\boxtimes)}$ ⓘ, Chan Wook Lee ⓘ, Seong Hyun Lim ⓘ, and Hyeong Suk Kim ⓘ

The University of Suwon, Hwaseong 18323, Republic of Korea
dgyoo411@suwon.ac.kr

Abstract. In this study, hydraulic procedures that can compensate problems in conventional model are newly suggested under seismic events in water distribution networks (WDNs). Detailed procedures for estimating the serviceability of WDNs using pressure driven analysis techniques, which can quantify available water supply under abnormal condition, are proposed. The methodologies are applied to Anytown network and get the result to compare with existing method. Proposed method leaded more appropriate hydraulic analysis results in abnormal situations such as earthquakes.

Keywords: Water distribution networks · Hydraulic analysis · Pressure driven analysis · Seismic events

1 Introduction

Water distribution networks (WDNs) are one of the social infrastructures, have functions to transport, distribute, and supply clean water, and are very complex connected systems consisting of pipes, tank, pumps, and valves. WDNs are large in scale and spatially widely distributed, and most of them are buried underground. Therefore, they are vulnerable to earthquake that can cause large damage.

Regarding the seismic reliability evaluation reflecting the hydraulic situations in WDNs, Wang (2006), Shi (2006), Yoo (2013), and Hou and Du (2014) conducted earthquake hazard quantification studies linked with hydraulic analysis on actual pipe networks. Yoo (2013) suggested REVAS.NET (Reliability EVAluation model of earthquake hazard for water supply NETwork), which is a model for assessment hydraulic serviceability based on structural failure. Yoo et al. (2016a) applied REVAS.NET to real WDNs, and it contains optimal pipe design module using Harmony Search Algorithm (HSA) (Yoo et al. 2016b).

However, studies related to the proposed earthquake hazard performance quantification used EPANET2 (Rossman 2000), a representative model of demand driven analysis (DDA). DDA performs hydraulic analysis that all the quantities of water supplied in the pipe network are considered as known values, and the quantities of water considered during hydraulic analysis are supplied without any shortages. However, existing model used EPANET2 (Rossman 2000), a representative model of

© Springer Nature Switzerland AG 2020
J. H. Kim et al. (Eds.): ICHSA 2019, AISC 1063, pp. 169–174, 2020.
https://doi.org/10.1007/978-3-030-31967-0_19

demand driven analysis (DDA). Therefore, DDA cannot directly simulate water leakage and breakage due to earthquake hazards. In realistic situation, failure degrades consumers' water availability due to water pressure drops.

The types of hydraulic analysis techniques of WDNs can be divided into as shown in Fig. 1 and suggested that in the case of abnormal situations such as earthquake hazards or areas where the leakage is large, since the supply and the usage are not the same, pressure driven analysis (PDA) that can consider changes in the supply and leakage quantities according to water pressure should be conducted.

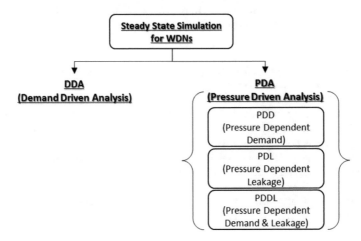

Fig. 1. Types of hydraulic analysis techniques for WDNs

A detailed processes are proposed for estimating the reliability of WSNs using a PDA that derives more appropriate hydraulic analysis results in abnormal situations such as earthquakes. In addition, reliability quantification factors such as system serviceability (S_s) and leakage ratio index (L_{ri}) are analyzed. Finally, the proposed methodologies are applied to representative WSNs and compared and evaluated with the results of REVAS.NET applied with DDA based model.

2   Methodology

2.1   REVAS.NET

REVAS.NET quantifies water supply serviceability through a series of processes: generation of stochastic seismic event, attenuation, determination of components' failure status (pipe, tank, and pump), and failure modeling based on hydraulic simulation tool. A Monte Carlo simulation (MCS) is used to quantify reasonable probabilistic seismic reliability. Figure 2 shows model construction of REVAS.NET.

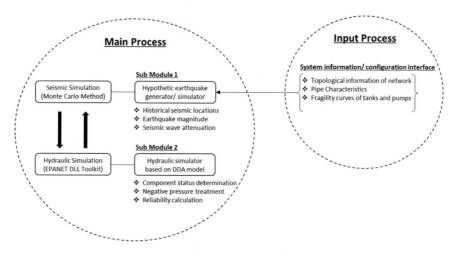

Fig. 2. Model construction of REVAS.NET

2.2 PDA Based Seismic Reliability Evaluation Model

Table 1 shows a comparison between the analysis of REVAS.NET and that of the proposed model when earthquake damage is presumed on the pipe line. Table 1 shows differences between the analysis of REVAS.NET and that of the proposed model when earthquake damage is defined on the pipe. In REVAS.NET, when leakage has occurred in the pipe line, the emitter coefficient is entered into the bottom node of the leakage pipe. However, in the proposed model, two coefficients (leakage coefficient/breakage coefficient) of the fixed and variable area discharge (FAVAD) expression proposed by May (1994)'s equation are directly entered and the leakage quantity of the pipe line is directly estimated.

Table 1. Evaluation indices for model assessment

Status of pipe	REVAS.NET	Proposed model
Leakage	The emitter coefficient is entered into the bottom node of the pipe where the leakage has occurred (single coefficient)	The C1 and C2 values of the FAVAD equation are entered into the pipe where the leakage has occurred (leakage coefficient/breakage coefficient)
Breakage	(1) During the simulation of hydraulic analysis, the pipe line state is set to "closed" to block flows (2) The emitter coefficient is entered into the top node of the broken pipe line	(1) During the simulation of hydraulic analysis, the pipe line state is set to "closed" to block flows (2) The FAVAD coefficient is entered into the broken pipe line
Hydraulic simulation technique	Quasi-PDA	Full-PDA

System serviceability (S_s) and leakage ratio index (L_{ri}) are set as performance indices in this simulation. S_s is an indicator intended to quantify the results of earthquake simulations through hydraulic analysis when earthquake simulations are implemented. S_s can calculate based on Eq. (1).

$$\text{System Serviceability } (S_s) = \frac{\sum Q_{avl,i}}{\sum Q_{inl,i}} \tag{1}$$

Where, $Q_{avl,i}$ is an available nodal demand at node i; $Q_{inl,i}$ is a required nodal demand at node i.

The second index is L_{ri}, which is a factor for evaluating the serviceability of the system. It is expressed as the ratio of the total required demand of the system to the leakage quantity of the entire pipe as shown in Eq. (2).

$$\text{Leakage Ratio Index } (L_{ri}) = \frac{\sum L_l}{\sum Q_{inl,i}} \tag{2}$$

Where, L_i is a leakage rate at pipe i; $Q_{inl,i}$ is a required nodal demand at node i.

3 Application Results

3.1 Application Network

The proposed method is applied to Anytown Network, a well-known benchmark network in the field of waterworks pipe networks. The demand required by the entire system is 17,640 GPM, and the total length of the pipe line is 267,000 ft. The pipe diameters are composed of form 6 to 30 in (Fig. 3).

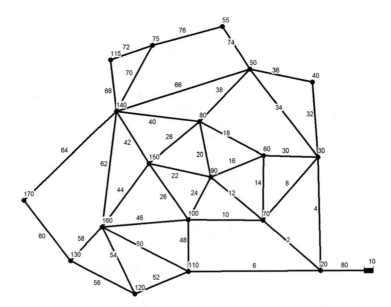

Fig. 3. Anytown network

3.2   Results of Scenarios

The six scenarios are set and their results were compared with REVAS.NET. In the Table 2, the number of pipes where leakage or breakage are shown depending on the scenario in which an earthquake of magnitude 4 occurred. In general, it can be predicted that the influence of pipe breakage is relatively large compare to leakage because the entire flow rate is lost.

Table 2. Scenarios depending on pipe status

Scenarios	Number of pipe		
	Normal	Leak	Breakage
S1	18	16	4
S3	16	19	3
S4	16	19	3
S7	18	17	3
S8	22	15	1
S10	10	27	1

Table 3 shows the comparison results of the proposed model and REVAS.NET. The results of a total of 6 scenarios showed that the average value of serviceability S_s is 0.821. This value means that if M4 earthquake occurs evenly in this system, water cannot be supplied for 17.9% of the total demand. The S_s value of REVAS.NET which is directly comparable is shown to be 0.294, indicating that the serviceability of the proposed model is about 2.8 times higher than that of REVAS.NET.

In case of L_{ri}, it can be thought to be the ratio of the flow rate lost in the pipe before being supplied to consumers due to earthquakes to the supply in normal situations. The average L_{ri} value calculated with the proposed model is 0.360, indicating that the flow rate corresponding to 36.0% of actual demand can be lost due to pipe leakage/breakage following earthquake hazards.

Table 3. Comparison results between proposed model and REVAS.NET

Scenarios	Proposed model						REVAS. NET
	Actual water supply (GPM)	Deficit (GPM)	Outflow (GPM)	Leakage (GPM)	S_s	L_{ri}	S_s
S1	12782	4858	20834	8051	0.725	0.456	0.202
S3	16004	1636	21705	5701	0.907	0.323	0.354
S4	16613	1027	24249	7636	0.942	0.433	0.246
S7	12989	4651	19756	6767	0.736	0.384	0.199
S8	11478	6162	17115	5637	0.651	0.320	0.305
S10	17017	623	21285	4269	0.965	0.242	0.459
Average	14481	3160	20824	6344	0.821	0.360	0.294

4 Conclusions

In the present study, detailed procedures for the estimation of the reliability of WDNs are proposed Using PDA technique. The developed methodologies can simulate many pipe leakage and breakage situations more realistically. The methodologies are applied to representative virtual pipe networks to review the models and new performance quantification indicators are additionally presented for analysis. Exist model, such as REVAS.NET showed a tendency to underestimate the reliability of the system, and such results are actually identified to be unrealistic. In addition, the two indicators proposed in the present study are identified to be mutually complementary when the supply flow rates are similar, and mainly usable as factors for evaluation of reliability from the viewpoints of users and suppliers, respectively. In particular, S_s, which is an existing system reliability evaluation factor, has a disadvantage of being unable to evaluate the water loss caused by earthquake hazards and can be utilized in judging whether suppliable from the viewpoint of users. L_{ri}, which is an index considering water loss, is a factor that indicates the degree of water loss due to earthquakes and can be used as a measure for determining whether to continue the operation of the system following the occurrence of earthquakes from the viewpoint of suppliers.

Acknowledgements. This work was supported by a grant from The National Research Foundation (NRF) of Korea, funded by the Korean government (Ministry of Science and ICT, MSIT) (No. NRF-2018R1C1B5046400).

References

Hou, B.W., Du, X.L.: Comparative study on hydraulic simulation of earthquake-damaged water distribution system. In: International Efforts in Lifeline Earthquake Engineering, pp. 113–120 (2014)

May, J.: Leakage pressure and control. In: BICS International Conference on Leakage Control, London, UK (1994)

Rossman, L.A.: EPANET 2: User's manual (2000)

Shi, P.: Seismic response modeling of water supply systems, Ph.D. thesis, Cornell University, Ithaca, NY, USA, 1 January 2006

Wang, Y.: Seismic performance evaluation of water supply systems, Ph.D. thesis, Cornell University, Ithaca, NY, USA, 1 January 2006

Yoo, D.G.: Seismic reliability assessment for water supply networks, Ph.D. thesis, Korea University, Republic of Korea (2013)

Yoo, D.G., Jung, D., Kang, D., Kim, J.H., Lansey, K.: Seismic hazard assessment model for urban water supply networks. J. Water Resour. Plan. Manag. ASCE **142**(2) (2016a). https://doi.org/10.1061/(ASCE)WR.1943-5452.0000584

Yoo, D.G., Kang, D., Kim, J.H.: Optimal design of water supply networks for enhancing seismic reliability. Reliab. Eng. Syst. Saf. **146**(2), 79–88 (2016b)

Research on an Evaluation of the Work Suitability in the New First-Tier Cities

Li Ma, Shi Liu$^{(\boxtimes)}$, Longhai Ai, and Yang Sun

Department of Computer Science, Inner Mongolia University, Hohhot, China
244046168@qq.com, liushi@imu.edu.cn

Abstract. It has become the new goal of urban development and the common demand of people to improve the city's level of employability and create a favorable environment for employment and entrepreneurship. Based on the functions of employment-friendly cities, this paper constructs an evaluation index system of five dimensions including population economy, life quality, social security, education culture and ecological environment, and adopts the entropy-TOPSIS model to evaluate the work suitability in the new first-tier cities, and tries to improve the entropy weight-TOPSIS model by using the vertical distance instead of the Euclidean distance. The new first-tier cities' work suitability rankings obtained by the two methods is similar, which indicates that the evaluation index system is reasonable and the evaluation results are credible.

Keywords: Evaluation of work suitability ·
Improved entropy-TOPSIS model · New first-tier cities

1 Introduction

Urban development is a symbol of the maturity and civilization of human society and an advanced form of human social life. The city is built for people's residence, and "livable" is the most primitive and basic appeal of cities. Therefore, in the process of urban development, the concept of "livable city" was first proposed for the evaluation of urban functional characteristics. Already in the 19th century, British scholar Howard proposed the concept of "garden city", which is considered to be the germination of modern thought of "livable city" [1]. The research on "livable cities" in China, by contrast, is relatively late. In 1989, scholar Wu published the book "Generalized Architecture", which started the research on the science of human living environment in China [2]. In 2005, the concept of "livable city" first appeared in the "Beijing City Master Plan" approved by the State Council. The employment-friendly city is a concept of urban form relative to "livable cities". "Livable" is the most basic functional feature of a city, and "work suitability" is a higher requirement for urban development. The relationship between urbanization level and industrial employment has become the focus of scholars at home and abroad. German scholar Liao studied the relationship between industrial agglomeration and urban formation and urbanization, and believed that "cities are point-shaped agglomerations of non-agricultural locations" [3]. Northam analyzed the development of many countries in the world, and concluded that the level of urbanization and the development of industrial structure are phased [4]. Kuznets believes that there is a clear constraint between

J. H. Kim et al. (Eds.): ICHSA 2019, AISC 1063, pp. 175–185, 2020.
https://doi.org/10.1007/978-3-030-31967-0_20

urbanization and industrial employment and further perfects the "Clark's theorem" [5]. Scholar Li studied and analyzed the relationship between urbanization and employment structure in China [6]. With the continuous improvement of urban functions, the construction of an employment-friendly city is crucial. In February 2019, the Central Committee of the Communist Party of China and the State Council issued the "Guangdong-Hong Kong-Macao Dawan District Development Plan", which proposed to build the area into a vibrant world-class city cluster and a high-quality living circle suitable for living, working and traveling. Obviously, a competitive world-class city should be a city that is ecologically healthy and livable with the citizens living and working in peace and contentment. It can attract the talents of all parties to fight here.

Through research literatures, it is found that compared with livable cities, there is little research on the employment-friendly city in the academic circles at present, and there is no unified definition of it. The concept of employment-friendly city appears more in government documents and news media. Employment-friendly city can be understood as a city that is more convenient for people to get a job and start a business at a certain level. Usually it refers to a city with a high level of economic development, which can provide sufficient employment and entrepreneurship opportunities, and guarantee people to get a generous return on employment. It has a favorable environment for employment, entrepreneurship and innovation. Therefore, it has strong attraction and agglomeration force to the population and excellent talents [7].

The new first-tier cities refer to those that can just be counted as first-tier cities and have great potential for development, but there is still a certain gap with first-tier cities. The list of new first-tier cities is determined according to a series of indicators such as political, economic and academic resources. The selection criteria and results are self-developed and selected by China Business Weekly. In 2018, the list of new first-tier cities includes Chengdu, Hangzhou, Chongqing, Wuhan, Suzhou, Xi'an, Tianjin, Nanjing, Zhengzhou, Changsha, Shenyang, Qingdao, Ningbo, Dongguan and Wuxi.

In recent years, the national urban development has become more balanced. Under the condition that population capacity and employment opportunities in the first-tier cities are becoming saturated, new first-tier cities have exploded with great development potential. In order to improve competitiveness, new first-tier cities have also optimized the living and professional development environment, providing better quality and conditions than first-tier cities in housing, employment, education, etc. And more talented people choose to work and live in new first-tier cities with the same development potential and lower settlement costs. Therefore, it has great theoretical and practical significance to evaluate and analyze the "suitable employment level" of new first-tier cities. This paper takes new first-tier cities in 2018 as the research object, builds an evaluation model by combining entropy weight method and TOPSIS, and evaluates the workability of each city. It is hoped to provide reference for the development of new first-tier cities and provide valuable reference for graduates' employment and entrepreneurship.

2 Construction of City Industrial Evaluation Index System

As the saying goes, "live and work in peace and contentment". During the research process, it is found that graduates will also consider settlement issues and choose a relatively livable city when choosing a city for employment and entrepreneurship. This also shows that in a certain degree of urban construction, work suitability and livability are mutually reinforcing. Therefore, the construction of the city's industrial evaluation index system should draw on the city's livability indicators.

However, "industry city" is different from livable cities. Livable cities refer to cities with social civilization, economic affluence, environmental grace, resource carrying capacity, low cost of life, and high public safety. The city's comprehensive livability index is above 80 and there are no negative conditions [8]. But the employment-friendly city refers to a city where people are engaged in various economic activities regardless of the natural environment or the social humanistic environment. Livable city emphasizes comfortable and convenient living environment; "Industry city" focus on providing more job opportunities and entrepreneurial conditions for people. In comparison, "industry city" has a higher level of livability and a wider coverage. In addition to being livable, it should also be conducive to the convergence of talents, technology and other high-quality resources.

Therefore, based on the functions of employment-friendly city and reading a large number of relevant literatures [9–11], an industrial index system is constructed from five dimensions of population economy, quality of life, social security, education and culture, and ecological environment to fully reflect the a city's work suitability, as shown in Table 1.

Table 1. City's industrial Evaluation Index System.

Primary indicators	Primary indicators
Population economy	Per capita disposable income of urban residents (RMB)
	Per capita GDP (RMB)
	Proportion of tertiary industry in GDP (%)
	Number of large-scale industrial enterprises (unit)
	Gross industrial product (billion RMB)
	Average annual salary of employees (RMB)
	Unemployment rate (%)
	Proportion of population aged 18–59 (%)
Life quality	Per capita housing area (m^2)
	Medical services per 10,000 people (person)
	Number of public transportation vehicles per 10,000 people (unit)
Social security	Urban endowment insurance coverage rate (%)
	Unemployment insurance coverage (%)
	Coverage rate of urban medical insurance (%)
	Proportion of public security expenditure in total fiscal expenditure (%)
	Proportion of social security and employment expenditure in total fiscal expenditure (%)

(*continued*)

Table 1. (*continued*)

Primary indicators	Primary indicators
Education culture	Proportion of R&D in GDP (%)
	Number of college graduates (person)
	Number of libraries and museums (unit)
	Number of universities (unit)
Ecological environment	Per capita park green space (m^2)
	Annual good air quality rate (%)
	PM2.5 annual average concentration ($\mu g/m^3$)
	Industrial sulfur dioxide emissions (ton)

3 Study on the Industrial Evaluation of New First-Tier Cities Based on Entropy-TOPSIS Model

3.1 Discussion on Practicality of Entropy-TOPSIS Model

This study evaluates the new first-tier city's work suitability, that is, evaluates multiple indicators in multiple cities, and finally ranks the evaluation results. The cities with higher rankings are more suitable for industry.

The entropy method [12] is an objective method of weighting, which uses the amount of information provided by the entropy of each indicator to determine the index weight. Using the principle of entropy to empower indicators, avoiding the subjectivity of expert scoring, reducing the workload of evaluation, and objectively and effectively reflecting the relative importance of each indicator [13]. The entropy method can avoid the interference of human factors and make the evaluation result more practical. By calculating the entropy value of each indicator, the amount of information of the indicator can be measured, so as to ensure that the established indicator can reflect most of the original information [14]. Therefore, comparing the subjective weighting methods such as AHP, the entropy method is used to weight the indicators based on objective data, avoiding the interference of subjective factors as much as possible, and making the weight determination more scientific and reasonable.

TOPSIS (Technique for Order Preference by Similarity to Ideal Solution) is an ordering method that approximates the ideal solution. It is a systematic evaluation method suitable for multi-index and multi-scheme decision analysis. It has clear ideas, simple calculation and strong operability, which is very suitable for this study. The traditional TOPSIS method calculates the weighted Euclidean distance between a scheme and the positive and negative ideal solutions, and obtains the close degree between the evaluation scheme and the positive ideal solution, which serves as the basis for evaluating the merits of each scheme [15]. However, in many evaluation schemes, the scheme that is closer to the positive ideal solution may also be closer to the negative ideal solution, so the evaluation results obtained cannot truly reflect the advantages and disadvantages of the schemes to be evaluated, making the evaluation results lack authenticity and rationality, which is the defect of using Euclidean distance [16].

In order to make up for the defects of the Euclidean distance in the traditional TOPSIS method, this paper uses the vertical distance instead of the Euclidean distance to improve, and calculates the vertical distance by orthogonal projection method to sort the evaluation scheme. The vertical distance is the plane where the positive ideal solution and the negative ideal solution are used as the normal vector, and the vertical normal vector is used to make two solutions, that is, the distance between the positive ideal solution and the negative ideal solution plane [17]. In all schemes, the scheme that is closer to the positive ideal solution and farther from the negative ideal solution is better.

Therefore, this paper uses the entropy-TOPSIS model and the improved entropy-TOPSIS model to study the industrial evaluation of new first-tier cities, in order to obtain objective and credible evaluation results.

3.2   Entropy-TOPSIS Model Evaluation Steps

An initial matrix is established for n evaluation indicators of m cities: $X = (x_{ij})_{m \times n}$, where x_{ij} represents the value of the j-th indicator of the i-th city $(i = 1, 2, \ldots, m; j = 1, 2, \ldots, n)$.

Step 1: Standardize decision matrix. In order to eliminate the impact of different indicators on the program, it is necessary to standardize the decision matrix and build a standardized matrix. The standardization process can use the following formula:

For the efficiency indicators which is the larger the better:

$$r_{ij} = \frac{x_{ij} - \min(x_{ij})}{\max(x_{ij}) - \min(x_{ij})}, i = 1, 2, \ldots, m; j = 1, 2, \ldots, n \tag{1}$$

For the cost indicators which is the smaller and better:

$$r_{ij} = \frac{\max(x_j) - x_{ij}}{\max(x_j) - \min(x_j)}, i = 1, 2, \ldots, m; j = 1, 2, \ldots, n \tag{2}$$

Step 2: Calculate the proportion of the i-th city in index j-th in this index:

$$p_{ij} = \frac{r_{ij}}{\sum_{i=1}^{m} r_{ij}}, i = 1, 2, \ldots, m; j = 1, 2, \ldots, n \tag{3}$$

Step 3: Calculate the entropy value of the j-th indicator e_j:

$$e_j = -\frac{1}{\ln(m)} \sum_{i=1}^{m} p_{ij} \ln(p_{ij}), j = 1, 2, \ldots, n \tag{4}$$

When $p_{ij} = 0$, $p_{ij} \ln(p_{ij}) = 0$.

Step 4: Calculate the difference coefficient g_j of the j-th indicator:

$$g_j = 1 - e_j, j = 1, 2, \ldots, n \tag{5}$$

For the j-th indicator, the greater difference in the indicator value, the more information reflected by the index.

Step 5: Calculate the entropy weight of each indicator:

$$w_j = \frac{g_j}{\sum\limits_{j=1}^{n} g_j}, j = 1, 2, \ldots, n \tag{6}$$

Step 6: Construct the weighted decision matrix:

$$v = (v_{ij})_{m \times n} = (w_j r_{ij})_{m \times n}, i = 1, 2, \ldots, m; j = 1, 2, \ldots, n \tag{7}$$

Step 7: Calculate the positive ideal solution and the negative ideal solution. According to the weighted decision matrix, the positive ideal solution S_j^+ and the negative ideal solution S_j^- of each scheme can be determined. S_j^+ is the optimal value of each index in the weighted decision matrix, and S_j^- is the worst value of each indicator.

Step 8-1: Calculate the Euclidean distance and sort:

$$Sd_i^+ = \left(\sum_{j=1}^{n} (S_j^+ - v_{ij})^2 \right)^{\frac{1}{2}}, i = 1, 2, \ldots, m \tag{8}$$

$$Sd_i^- = \left(\sum_{j=1}^{n} (S_j^- - v_{ij})^2 \right)^{\frac{1}{2}}, i = 1, 2, \ldots, m \tag{9}$$

$$\delta_i = \frac{Sd_i^-}{Sd_i^- + Sd_i^+}, i = 1, 2, \ldots, m \tag{10}$$

Sorting according to the size of δ_i, the larger the δ_i, the better the industry's suitability.

Step 8-2: Calculate the vertical distance and sort. In order to simplify the calculation, the positive ideal solution is shifted to the origin of coordinates, and the translated evaluation matrix M is obtained:

$$M = \left(v_{ij} - S_j^+ \right)_{m \times n} = (m_{ij})_{m \times n} \tag{11}$$

Calculate the vertical distance from evaluation scheme i to positive ideal solution:

$$d_i = |(p - q) \bullet (p - M_i)| = |q \bullet M_i| \tag{12}$$

Where p and q are the positive and negative ideal solutions after translation, and p is a zero vector, M_i is the i-th row vector of the evaluation matrix M after translation. According to the size of d_i, the smaller the value of d_i, the better the city's suitability.

4 Empirical Research

4.1 Source of Indicator Data

The new first-tier cities are selected by China Business Weekly. In 2018, the "new first-tier cities" include Chengdu, Hangzhou, Chongqing, Wuhan, Suzhou, Xi'an, Tianjin, Nanjing, Zhengzhou, Changsha, Shenyang, Qingdao, Ningbo, Dongguan and Wuxi.

In order to understand the latest suitable conditions of various cities and make contributions to urban construction, this paper adopts the data of 2018. The indicator data comes from the 2018 statistical yearbooks, the 2018 statistical bulletins and the relevant government official website information.

4.2 Model Solution

In accordance with the evaluation steps in Sect. 3.2, this paper uses MATLAB2016a to write the entropy weight-TOPSIS program with Euclidean distance and vertical distance respectively for data processing [18], which needs to pay attention on that the secondary indicators unemployment rate (%), PM2.5 annual average concentration ($\mu g/m^3$), industrial sulfur dioxide emissions (ton) are cost-based indicators, and the remaining secondary indicators are benefit-based indicators. By running the program, the weight of each index is shown in Table 2. The ranking and the close degree between each city and positive ideal solution calculated by the two methods are shown in Figs. 1 and 2 respectively.

4.3 Analysis of Results

Population Economy is the Main Factor Affecting the City's Work Suitability.
According to Table 2, the weight of the population economy in the primary indicators (0.2734) is the largest, while the per capita disposable income of urban residents in the population economy (0.253) is the highest. Therefore, for the industrial evaluation of new first-tier cities, the level of economic development is still the main factor. The level of economic development is positively related to the ability to pool high-quality resources. The higher the level of economic development, the stronger the attraction and concentration of capital, technology, talents and other high-quality resources. Economic

Table 2. Evaluation index weight table of new first-tier city's work suitability.

Primary indicators	The weight	Secondary indicators	The weight
Population economy	0.2734	Per capita disposable income of urban residents (RMB)	0.253
		Per capita GDP (RMB)	0.0301
		Proportion of tertiary industry in GDP (%)	0.0280
		Number of large-scale industrial enterprises (unit)	0.0512
		Gross industrial product (billion RMB)	0.0509
		Average annual salary of employees (RMB)	0.0309
		Unemployment rate (%)	0.0338
		Proportion of population aged 18–59 (%)	0.0232
Life quality	0.1512	Per capita housing area (m^2)	0.0405
		Medical services per 10,000 people (person)	0.0819
		Number of public transportation vehicles per 10,000 people (unit)	0.2888
Social security	0.2434	Urban endowment insurance coverage rate (%)	0.0395
		Unemployment insurance coverage (%)	0.0519
		Coverage rate of urban medical insurance (%)	0.0222
		Proportion of public security expenditure in total fiscal expenditure (%)	0.0580
		Proportion of social security and employment expenditure in total fiscal expenditure (%)	0.0718
Education culture	0.2472	Proportion of R&D in GDP (%)	0.0400
		Number of college graduates (person)	0.0430
		Number of libraries and museums (unit)	0.1302
		Number of universities (unit)	0.0340
Ecological environment	0.0852	Per capita park green space (m^2)	0.0273
		Annual good air quality rate (%)	0.0235
		PM2.5 annual average concentration (μg/m^3)	0.0206
		Industrial sulfur dioxide emissions (ton)	0.0138

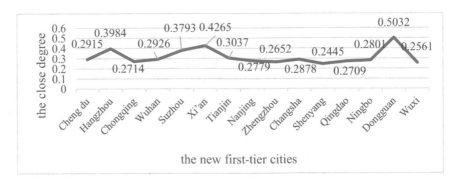

Fig. 1. The line chart of new first-tier cities' city's work suitability (using Euclidean distance)

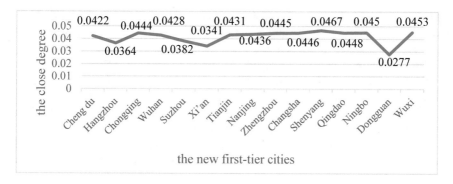

Fig. 2. The line chart of new first-tier cities' city's work suitability (using vertical distance)

agglomeration drives population agglomeration, especially to attract more young labor force, and the agglomeration of young labor force in turn promotes economic agglomeration. In other words, the stronger the economic strength of the city, the more industrial enterprises above designated size, the more job opportunities they can provide, the more attractive the young workforce, and the higher the industrial production. Therefore, the city's economic development level represents the comprehensive strength of a city and is an important indicator for evaluating the city's work suitability.

Education Culture is a Secondary Factor Affecting the City's Work Suitability.
As shown in Table 2, the index of educational culture has a high weight. The education culture of a city is the soft power of a city. It advocates innovation and respects knowledge, and provides an atmosphere and space conducive to personal career growth. It can stimulate people's enterprising and innovative spirit and enhance the sense of identity of residents. Libraries and museums are an important part of urban culture, and the number of which can reflect the strength of urban education and culture.

Urban Quality is an Important Factor Affecting the City's Work Suitability. The quality of a city is an important factor that affects the comfort of people living and working in the city. The social security system is one of the most important social and economic systems in modern countries, which is used to ensure the basic survival and living needs of the whole society. A sound social security system can guarantee people's life, make them more secure and practical in work, improve the happiness of urban residents, and improve the city's industrial suitability. At the same time, modern people pay more and more attention to the quality of life and pursue a higher quality of life. Housing, medical care, and transportation problems are closely related to everyone's life, directly affecting people's quality of life, and thus affecting city livability and suitability. A city that provides people with comfortable, convenient and high-quality living, makes people work comfortably, and makes the urban population more attractive. In addition, the ecological environment cannot be ignored. A good ecological environment can enable residents to enjoy an excellent urban living environment, which is conducive to physical and mental health, and makes people more willing to settle and work here.

Analysis of the Ranking Results of the New First-Tier Cities' Work Suitability. As shown in Figs. 1 and 2, the results of the new first-tier cities' work suitability ranking calculated by Euclidean distance are: Dongguan, Xi'an, Hangzhou, Suzhou, Tianjin, Wuhan, Chengdu, Changsha, Ningbo, Nanjing, Chongqing, Qingdao, Zhengzhou, Wuxi, Shenyang. And the results of the new first-tier cities' work suitability ranking calculated by the vertical distance are: Dongguan, Xi'an, Hangzhou, Suzhou, Chengdu, Wuhan, Tianjin, Nanjing, Chongqing, Zhengzhou, Changsha, Qingdao, Ningbo, Wuxi, Shenyang. From the ranking results obtained by the two methods, the 15 new first-tier cities have similar employment ranking, which also indicates that the evaluation index system constructed in this paper is reasonable and the evaluation results are credible.

5 Conclusion

Based on the index of urban livability and functions of employment-friendly city, this paper constructs an evaluation index system including the five dimensions of population economy, life quality, social security, education culture and ecological environment. This paper provides a basis for the evaluation and construction of new first-tier city's work suitability. In this paper, entropy-TOPSIS method is adopted to combine the objective weighting of entropy method with the multi-attribute decision ranking of TOPSIS method, so as to avoid the subjectivity of artificial weight determination. The TOPSIS method is improved by replacing Euclidean distance with vertical distance in order to make the evaluation result scientific and reasonable. Therefore, the discussion and research in this paper is of positive significance, which is conducive to promoting the construction of urban work suitability.

References

1. Zhou, C., Deng, H.: A review of the theory of livable cities abroad. J. Hefei Univ. Technol. (Soc. Sci. Edn.) **25**(4), 62–67 (2011)
2. Wu, L., Liu, Y.: Evaluation of livability of major cities in China. J. Shaanxi Normal Univ. **24**(2), 226–229 (2010)
3. Liao, S.: Economic Space Order. Shouli Wang Translated. Commercial Press, Beijing (2010)
4. Lewis, W.A.: Unlimited labour: further notes. Manch. Sch. **26**(1), 1–32 (2010)
5. Simon, K.: Modern economic growth: rate, structure, and spread Canadian. J. Econ. Polit. Sci. **33**(33), 475–476 (1967)
6. Li, L., Yu, F.: Statistical test of the dependence relationship between urbanization and employment structure in China. Stat. Decis. **13**, 84–85 (2006)
7. Li, S., Chen, X.: There is no "Yiye", how come "livable"? – Analysis of the construction of livable living city in Chengdu. J. Chengdu Univ. (Soc. Sci. Edn.) **6**, 39–45 (2017)
8. Luo, Y., Ren, Z., Zhen, F.: Research on livable city science evaluation index system. China Urban Science Research Association (2007)
9. Zhang, Y., Zhu, J.: Fuzzy comprehensive evaluation of livable city based on TOPSIS. J. Harbin Univ. Commer. (Nat. Sci. Edn.) **34**(4), 508–512 (2018)

10. Zhang, C., Wang, J.: Urban residential evaluation of Huaihai economic zone. J. Changchun Univ. Technol. **39**(3), 296–300 (2018)
11. Sofeska, E.: Understanding the livability in a city through smart solutions and urban planning toward developing sustainable livable future of the city of Skopje. Procedia Environ. Sci. **37**, 442–453 (2017)
12. Zhu, X., Wei, G.: Discussion on the excellent standard of dimensionless method in entropy method. Stat. Decis. **2**, 12–15 (2015)
13. Zheng, X., Du, J., Su, Y.: Evaluation of urban sustainable development based on improved entropy weight method——taking Harbin city as an example. J. Civ. Eng. Manag. **35**(04), 65–71 (2018)
14. Zhang, S., Zhang, M., Chi, G.: Science and technology evaluation model based on entropy weight method and its empirical research. J. Manag. **1**, 34–42 (2010)
15. Chen, L., Wang, Y.: Research on decision-making method based on entropy weight coefficient and TOPSIS integrated evaluation. Control Decis. **18**(4), 456–459 (2003)
16. Qi, Y., Li, C., Wang, X., Fan, Z.: Evaluation of DDoS defense measures for improving TOPSIS method. J. Weapon Equip. Eng. **40**(2), 158–162 (2019)
17. Liu, B., Zhang, Q., Wang, H.: Model and application research of water resource right allocation based on improved TOPSIS method. Groundwater **40**(6), 189–191 (2008)
18. He, F.: Comprehensive Evaluation Method MATLAB Implementation, pp. 321–332. China Social Sciences Press, Beijing (2010)

Optimal Path Planning in Environments with Static Obstacles by Harmony Search Algorithm

Eva Tuba, Ivana Strumberger, Nebojsa Bacanin, and Milan Tuba$^{(\boxtimes)}$

Singidunum University, Danijelova 32, 11000 Belgrade, Serbia
{etuba,tuba}@ieee.org, {istrumberger,nbacanin}@singidunum.ac.rs

Abstract. Path planning represents an important optimization problem that need to be solved in various applications. It is a hard optimization problem thus deterministic algorithms are not usable but it can be tackled by stochastic population based metaheuristics such as swarm intelligence algorithms. In this paper we adopted and adjusted harmony search algorithm for the path planning problem in environment with static obstacles and danger zones. Objective function includes path length and safety degree. The proposed method was tested on standard benchmark examples from literature. Simulation results show that our proposed model produces better and more consistent results in spite of its simplicity.

Keywords: Path planning · Optimization · Swarm intelligence · Harmony search algorithm

1 Introduction

Path planning is challenging problem where the optimal path between two points should be found [5,11] and it appears in various applications. It considers path planning for various objects in different environments. It can refers to problems such as two or three dimensional path planning for underwater automated vehicles, mobile sensors networks, unmanned aerial vehicles, robots in radioactive or unknown and uncertain environments with static or dynamic obstacles which include obstacles with uncertain position, moving or non-permanent ones such as fire, landmines, radioactive zones, etc. Based on the nature of environment and the applications, the optimal path can be considered if it is the shortest or smoothest, if the path avoids static and/or dynamic obstacles or avoids danger zones, etc. The path planning represents an old problem but without the perfect solution thus various research papers propose different methods for solving it.

In general, all proposed methods for solving path planning problem can be divided into two classes: classical and heuristic solutions. Classical solutions such as roadmaps, cell decomposition or mathematical programming, are with high computational cost and they have tendency to get stuck in local optima.

© Springer Nature Switzerland AG 2020
J. H. Kim et al. (Eds.): ICHSA 2019, AISC 1063, pp. 186–193, 2020.
https://doi.org/10.1007/978-3-030-31967-0_21

The path planning problem is an NP-hard problem thus heuristic approaches proved to be very successful. Current majority of path planning methods are using heuristics.

Nowadays, nature inspired metaheuristic algorithms are widely used for solving various hard optimization problems. Nature inspired algorithms are based on different phenomena from nature used as an inspiration for creating optimization algorithms. Swarm intelligence algorithms, a class of nature inspired algorithms, represent population based optimization metaheuristics. They use swarm of simple agents that can perform rather simple operations and they exchange informations between themselves and by that, swarm can exhibit remarkable intelligence. One of the oldest swarm intelligence algorithms are particle swarm optimization (PSO) [8] and ant colony optimization (ACO) [7]. Nowadays swarm intelligence algorithms have been widely used for solving various hard optimization problems in image processing [15,17,18], wireless sensor networks [3,13], machine learning [14,16], etc. They have been also applied for path planning problem [1,4,12].

Before using any optimization method, it is necessary to precisely define the objective function and a mathematical representation of a path. There are numerous objective functions and path representations used in the literature. In this paper we used a model where path is built form starting point to the target without possibility of moving back toward the start point, i.e. movement forward is ensured. Depending on the environment and demands, path planning problem can be defined as a single or multiobjective optimization problem. In this paper, objective function was defined by the degree of collision with static obstacles and path length. For solving described optimization problem, we adjusted harmony search algorithm.

The rest of this paper is organized in a following way. Description of the path representation and objective function is presented in Sect. 2. Adjusted harmony search algorithm is described in Sect. 3. Simulation results are presented in Sect. 4. Conclusion and plans for further work are given in the Sect. 5.

2 Path Representation and Mathematical Model

The path planning refers to a search for the optimal way to connect the starting point (S) and the target point (T) while satisfying certain demands. In this paper, the optimal path should avoid static obstacles thus the goal is to find the shortest collision free path.

Frequently used method for path representation is to use sequence of points and the path connects these points by straight line. Additionally, turn backs are not allowed, i.e. each point of the is closer to the target point than the previous one [2,6,10,12]. Path is represented by n points between the start and target point where each of them lies on the lines orthogonal start-target line. Mathematical definition of the described path is as follows.

Firstly, the x-axis of the used coordinate system is the line that connects the start (S) and target (T) points, i.e. S has coordinates (0,0). Transformation to this coordinate system is achieved by:

$$
\begin{bmatrix} x' \\ y' \end{bmatrix} = \begin{bmatrix} \cos\phi & -\sin\phi \\ \sin\phi & \cos\phi \end{bmatrix} \cdot \begin{bmatrix} x \\ y \end{bmatrix} + \begin{bmatrix} x_s \\ y_s \end{bmatrix},
\tag{1}
$$

where (x, y) and (x', y') are the original and transformed coordinates, respectively, (x_s, y_s) are coordinates of the S and θ represents anti-clockwise angle between x-axis and the vector $\overrightarrow{ST}$:

$$
\phi = \arcsin \frac{|y_T - y_S|}{||\overrightarrow{ST}||}.
\tag{2}
$$

Path points lies on the equidistant lines orthogonal to the $\overrightarrow{ST}$ (Fig. 1).

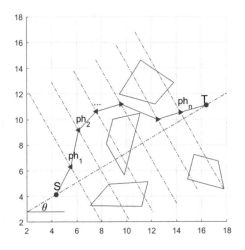

Fig. 1. Path representation

In this path representation, x coordinates of the path points ph_1, ph_2, ..., ph_n are known and y coordinates should be determined. The full path is defined as:

$$
Path = (S, ph_1, ph_2, \ldots, ph_n, T)
\tag{3}
$$

In order to calculate the objective function, the path length should be calculated and check is there are collisions. The path length is the sum of distances between each neighborhood path points:

$$
L(PH) = \sum_{i=0}^{n} d(ph_i, ph_{i+1})
\tag{4}
$$

where ph_0 is S, ph_{n+1} is T, $d(\cdot)$ is the Euclidean distance.

In objective function, we included penalty for paths that goes through obstacles. In described model, path planning represents minimization optimization problem where the objective function is defines as follows:

$$F(PH) = L(PH) + c * penalty \tag{5}$$

where c is the number of obstacles that cross with the path PH and $penalty > 0$ is the constant.

3   Harmony Search Algorithm

Harmony search (HSA) is an optimization approach that belongs to the group of so called phenomenon-mimicking metaheuristics. The HSA was first introduced in 2001 by Geem, Kim and Loganthan [9]. It was inspired by the improvisation process of jazz musicians.

For successfully implementation of a metaheuristics approach, idealization rules, that approximate real world system, should be applied. In the case of HSA algorithm, the improvisation process by a skilled musical is simplified by introducing three possible choices: musical may play any famous piece of music from her or his memory (pitches series in harmony), musician may play music that is similar to a known master piece, in which case she or he slightly adjusts the pitch or musician may compose and play pseudo-random music. Based on these rules, three basic components of the HSA metaheuristics are: utilization of harmony memory, pitch adjusting and randomization [9].

First HSA's component, utilization of harmony memory, is very significant due to the fact that it ensures that the best harmonies will propagate to the new harmony memory. The process that is carried by this component is similar to selecting the individuals with the highest fitness in genetic algorithm.

Utilization of harmony memory component uses harmony memory accepting parameter (or considering rate) $r_{accept} \in [0, 1]$. This parameter controls the balance between exploitation and exploration in the HSA's search process. If the value of considering rate parameter is too low, only few best solutions (harmonies) from the search space are selected, and the algorithm may converge slowly to the optimal region of the search space. Otherwise, if the value of this parameter is too high (near to 1), almost all solutions (harmonies) are used in the harmony memory, and other harmonies may not be explored well. This potentially may lead to the situation in which the search process is being trapped into one of the suboptimal regions of the search space. Due to these reasons, in typical implementations, the value for the r_{accept} is roughly between 0.7 and 0.95 [20].

The second component of the HSA metaheuristics, the pitch adjustment depends on the pitch bandwidth b_{range} and on a pitch adjusting rate r_{pa}. Pitch adjusting rate is used for generating new solutions in the neighborhood of current solutions, thus it is used for the purpose of exploitation process. In practice, the pitch is being adjusted linearly by using the following equation:

$$x_{new} = x_{old} + b_{range} \cdot \varepsilon, \tag{6}$$

where x_{old} represents existing solution (pitch) that resides in the population (harmony memory), x_{new} denotes novel generated solution, while the ε is pseudo-random number in the range $[0, 1]$.

Moreover, in most implementations, a pitch adjusting rate r_{pa} is also applied for the purpose of controlling the degree of the search process adjustments. It is important to choose right values for the r_{pa} and for the b_{range} parameters. If both these values are too low, a convergence of the HSA may be reduced, because of the limitation in the exploitation process to only small part of the whole search space. In the opposite case, if these values are too high, solutions may be scattered randomly in the search space, as in the case of random search methods. In most implementations, the value for the r_{pa} is typically between 0.1 and 0.5.

Algorithm 1. HSA pseudo-code

1: Generate initial pseudo-random solution population (harmonics), set iteration counter t to 1, and the value for maximum iteration number parameter $MaxIter$
2: Set values for the pitch adjusting rate (r_{pa}), pitch limit and bandwidth
3: Set value for the harmony memory accepting rate (r_{accept})
4: **while** $t < MaxIter$ **do**
5:    Generate new solutions (harmonics) by adjusting accepting best solutions
6:    Adjust pitch to get new solutions
7:    **if** $rand > r_{accept}$ **then**
8:       Choose randomly existing solution
9:    **end if**
10:   **if** $rand > r_{pa}$ **then**
11:      Adjust the pitch randomly within limits
12:   **else**
13:      Generate new solutions by applying randomization
14:   **end if**
15:   Accept new solution if it is better than the old one
16: **end while**
17: Find the best solution from the population

Finally, the third component of the HSA is randomization. The purpose of this component is to increase the diversity of the solutions in the population. The probability of randomization is expressed by the following equation:

$$P_{random} = 1 - r_{accept} \tag{7}$$

The actual probability of adjusting pitches is expressed as:

$$P_{pitch} = r_{accept} \cdot r_{pa} \tag{8}$$

Taking into account all three components, the pseudo-code of the HSA meta-heuristics can be summarized in the Algorithm 1.

In this paper, we used HSA algorithm for finding the optimal path for a model where y-coordinates need to be determined. The optimal path is defined by Eq.(5).

4   Simulation Results

All simulations in this paper were preformed on the platform with Intel ® Core™ i7-3770K CPU at 4 GHz, 8 GB RAM and Windows 10 Professional OS.

Parameters for the harmony search algorithm were empirically. Population size was set to $HM == 20$. The pitch adjusting rate r_{pa} was 0.3, memory accepting parameter r_{accept} was 0.85 while pitch bandwidth b_{range} was 0.1. The maximal iteration number was 100.

The quality of the proposed HSA method was tested by comparing the obtained results with the results presented in [19]. In [19] firefly algorithm (FA) was used for minimizing the path length by determining the control points of a Bezier curve. They tested the proposed FA method in four different environments. Exact positions of the obstacles placed in the fields of size 15 × 15 are given in [19]. Starting and target positions were S(5,5) and T(15,15). The results obtained in [19] were compared to genetic algorithm (GA) and particle swarm optimization with adaptive inertia weight (PSO).

Comparison of the average path lengths reported in [19] and obtained by our proposed HSA method is presented in Table 1. We average path length along with the standard deviation. For each test environment we started our method 30 times. Number of path points was 10 the first two test environments while for the third and the fourth test example, we set the number of path points to 15 since they are more complex based on the number of obstacles and the path that should be created.

Table 1. Comparison of the path lengths obtained by different methods

No.	GA	PSO	FA	HSA
P1	17.30	17.13	17.44	**16.03**±1.47
P2	19.07	17.85	17.96	**15.48**±1.50
P3	18.23	17.76	18.31	**16.99**±0.97
P4	19.83	19.36	20.62	**18.96**±0.23

Based on the results presented in Table 1, we can conclude that our proposed HSA method out performed all methods presented in [19]. Since the path model in [19] included Bezier curves, there were unnecessary turns in the founded paths. Paths obtained by our proposed method for all four environments are presented in Fig. 2.

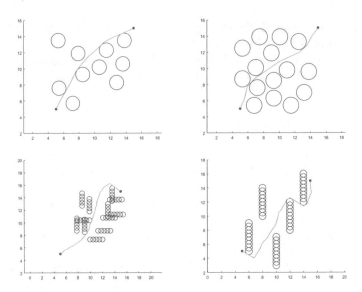

Fig. 2. The optimal path found by our proposed method for test environments used in [19]

5   Conclusion

In this paper we considered path planning problem. The proposed path model defines the optimal path as the shortest collision free path in environments with static obstacles. Adjusted harmony search algorithm was used for finding the optimal path. Simulation results showed that our proposed method is efficient for path planning problem and better compared to other state-of-the-art methods. Future work can include three dimensional environments, dynamic obstacles, danger zones, etc.

Acknowledgment. This research is supported by Ministry of Education, Science and Technological Development of Republic of Serbia, Grant No. III-44006.

References

1. Alam, M.S., Rafique, M.U., Khan, M.U.: Mobile robot path planning in static environments using particle swarm optimization. Int. J. Comput. Sci. Electron. Eng. (IJCSEE) **3**, 253–257 (2015)
2. Alam, M.S., Rafique, M.U.: Mobile robot path planning in environments cluttered with non-convex obstacles using particle swarm optimization. In: International Conference on Control, Automation and Robotics (ICCAR), pp. 32–36. IEEE (2015)
3. Chen, J., Cheng, S., Chen, Y., Xie, Y., Shi, Y.: Enhanced brain storm optimization algorithm for wireless sensor networks deployment. In: International Conference in Swarm Intelligence, pp. 373–381. Springer (2015)

4. Contreras-Cruz, M.A., Ayala-Ramirez, V., Hernandez-Belmonte, U.H.: Mobile robot path planning using artificial bee colony and evolutionary programming. Appl. Soft Comput. **30**, 319–328 (2015)
5. Delmerico, J., Mueggler, E., Nitsch, J., Scaramuzza, D.: Active autonomous aerial exploration for ground robot path planning. IEEE Robot. Autom. Lett. **2**(2), 664–671 (2017)
6. Dolicanin, E., Fetahovic, I., Tuba, E., Capor-Hrosik, R., Tuba, M.: Unmanned combat aerial vehicle path planning by brain storm optimization algorithm. Stud. Inf. Control **27**(1), 15–24 (2018)
7. Dorigo, M., Gambardella, L.M.: Ant colonies for the travelling salesman problem. Biosystems **43**(2), 73–81 (1997)
8. Eberhart, R., Kennedy, J.: A new optimizer using particle swarm theory. In: Proceedings of the Sixth International Symposium on Micro Machine and Human Science, pp. 39–43. IEEE (1995)
9. Geem, Z., Kim, J., Loganathan, G.: A new heuristic optimization algorithm: harmony search. Simulation **76**(2), 60–68 (2001)
10. Li, B., Gong, L.g., Yang, W.l.: An improved artificial bee colony algorithm based on balance-evolution strategy for unmanned combat aerial vehicle path planning. Sci. World J. **2014**, 1–10 (2014). Article ID 232704
11. Liu, J., Yang, J., Liu, H., Tian, X., Gao, M.: An improved ant colony algorithm for robot path planning. Soft Comput. **21**(19), 5829–5839 (2017)
12. Liu, Y., Zhang, X., Guan, X., Delahaye, D.: Adaptive sensitivity decision based path planning algorithm for unmanned aerial vehicle with improved particle swarm optimization. Aerosp. Sci. Technol. **58**, 92–102 (2016)
13. Tuba, E., Tuba, M., Beko, M.: Node localization in ad hoc wireless sensor networks using fireworks algorithm. In: 5th International Conference on Multimedia Computing and Systems (ICMCS), pp. 223–229. IEEE (2016)
14. Tuba, E., Tuba, M., Beko, M.: Support vector machine parameters optimization by enhanced fireworks algorithm. In: International Conference in Swarm Intelligence, LNCS, vol. 9712, pp. 526–534. Springer (2016)
15. Tuba, E., Tuba, M., Dolicanin, E.: Adjusted fireworks algorithm applied to retinal image registration. Stud. Inf. Control **26**(1), 33–42 (2017)
16. Tuba, E., Tuba, M., Simian, D.: Handwritten digit recognition by support vector machine optimized by bat algorithm. In: Computer Science Research Note: International Conference in Central Europe on Computer Graphics, Visualization and Computer Vision, pp. 369–376 (2016)
17. Tuba, E., Tuba, M., Simian, D., Jovanovic, R.: JPEG quantization table optimization by guided fireworks algorithm. In: IWCIA: Combinatorial Image Analysis, vol. 10256, pp. 294–307. Springer (2017)
18. Tuba, M., Bacanin, N.: JPEG quantization tables selection by the firefly algorithm. In: International Conference on Multimedia Computing and Systems (ICMCS), pp. 153–158. IEEE (2014)
19. Xianxiang, W., Yan, M., Juan, W.: An improved path planning approach based on particle swarm optimization. In: 11th International Conference on Hybrid Intelligent Systems (HIS), pp. 157–161. IEEE (2011)
20. Yang, X.S.: Harmony Search as a Metaheuristic Algorithm, pp. 1–14. Springer, Heidelberg (2009). https://doi.org/10.1007/978-3-642-00185-7_1

Online Feature Selection via Deep Reconstruction Network

Johan Holmberg and Ning Xiong[(✉)]

Mälardalen University, Västerås, Sweden
{johan.holmberg,ning.xiong}@mdh.se

Abstract. This paper addresses the feature selection problems in the setting of online learning of data streams. Typically this setting imposes restrictions on computational resources (memory, processing) as well as storage capacity, since instances of streaming data arrive with high speed and with no possibility to store data for later offline processing. Feature selection can be particularly beneficial here to selectively process parts of the data by reducing the data dimensionality. However selecting a subset of features may lead to permanently ruling out the possibilities of using discarded dimensions. This will cause a problem in the cases of feature drift in which data importance on individual dimensions changes with time. This paper proposes a new method of online feature selection to deal with drifting features in non-stationary data streams. The core of the proposed method lies in deep reconstruction networks that are continuously updated with incoming data instances. These networks can be used to not only detect the point of change with feature drift but also dynamically rank the importance of features for feature selection in an online manner. The efficacy of our work has been demonstrated by the results of experiments based on the MNIST database.

Keywords: Non-stationary · Data streams · Online feature selection

1 Introduction

A particular challenge in learning systems is the curse of dimensionality. High data dimensionality can adversely impact the performance of learning algorithms. It can also lead to the problem of overfitting which in turn can cause sub-optimal accuracy for prediction [11]. A common approach to reduce dimensionality is feature selection.

A challenge with non-stationary data streams in relation to feature selection is feature drift. Feature drift occurs when features selected for learning, either cease to be or become important for the learning task mid-stream [1]. This issue arises most commonly in the learning tasks of classification, yet it presents the importance whenever feature selection is required in a dynamic setting.

This paper proposes a new method of online feature selection to deal with drifting features in non-stationary data streams. The core of the proposed

© Springer Nature Switzerland AG 2020
J. H. Kim et al. (Eds.): ICHSA 2019, AISC 1063, pp. 194–201, 2020.
https://doi.org/10.1007/978-3-030-31967-0_22

method lies in a deep reconstruction network that is continuously updated with incoming data instances. This network can be used to not only detect the point of change with feature drift but also dynamically rank the importance of features for feature selection in an online manner. The efficacy of our work has been demonstrated by the results of experiments based on the MNIST database.

2   Related Work

2.1   Non Stationary Data Stream Learning

Adaptive learning in stationary data streams has been addressed in different ways in previous research. In terms of approaches, one finds the broad separation between active and passive methods [3]. In active approaches, typically a change point is sought to determine when adaptation should be applied. In passive approaches, rather than attempting to find a change point for adaptation (which in itself implies a cost) the problem of drift is indirectly tackled by continuous model updates.

Randomness/Combinatorics. Typically these methods rely on ensembles, where each algorithm is fed its own separate subset of features. The learning system as a whole could be able to adapt to drifting features, although there is no explicit mechanism that deals with feature drift.

Streaming stacking is a model which employs an ensemble of decision tress learning from distinct subsets of features. Streaming random forests and Random rules are two methods that use combinatorics and also randomness to address the issue [1].

Windowing. Sliding window based algorithms use windows of recent data instances to determine the level of drifting that has occurred. This makes it possible to adapt offline non streaming algorithms to the setting of online incremental learning. CVFDT [6] is a Hoefffding Tree learning algorithm which keeps statistics to enable a consistent model given a sliding window. HEFT-Stream [12] employs an ensemble of classifiers to select relevant features.

Deep Learning of Non-stationary Data. In [2] an approach is presented where a Deep Belief Network (DBM), called Adaptive DBN is used in a non-stationary environment for classification. It trains new generations of itself based on new data but also generates samples from it's predecessor to be able to learn from old training data without storing historical samples. Another deep learning approach for non-stationary data is presented in [4], where the reconstruction errors on new data are improved by modification of the neural network topology. By introducing new neurons via the proposed algorithm Neurogenesis Deep Learning (NDL) and training the new network by a replay mechanism, the network can learn to reconstruct new data.

2.2   Feature Selection

Broadly there are three classes of approaches to feature selection, filter, wrapper and embedded methods [11]. Filter methods try to statistically select features based on a criteria, while quick and easy to employ, there are drawbacks. Wrapper methods attempt to recreate a full model for evaluation of a feature subset. Embedded feature selection and/or feature importance ranking have received much attention [5, 10, 13], where linear approaches or single layer neural networks have been used to reconstruct the data. Particularly the Sparse learning approach has shown its success to rank and select the most important features. Nevertheless all the aforementioned works have focus on an offline learning paradigm.

3   Proposed Method

The proposed method for Online Feature Selection via deep reconstruction is described here. Section 3.1 describes how changes of features are detected actively. The feature selection method is described in Sect. 3.2. Section 3.3 describes the ensemble-like set-up and the policy for *active/passive* change adaptation.

3.1   The Feature Monitoring Approach

For the setting of unsupervised data stream learning with drifting features, we propose the approach referred to as Feature Monitoring (FM). The FM approach detects changes using the reconstruction error, and also provides direct guidance to downstream learning systems about what has changed in terms of feature importance to improve adaptations.

It is assumed that there is a data generating process $\mathcal{P}(x)$. We monitor the incoming features $p_t(x)$ to detect change, where $p_t(x)$ represents $\mathcal{P}(x)$. The Feature Monitoring approach uses a reconstruction component $\mathcal{R}$ which employs non-linear transformations to to project inputs to a curved manifold and then reconstruct them. This makes the solution more general compared to methods such as principal components analysis (PCA, [8]) or other linear methods. Once $\mathcal{R}$ is used to calculate the reconstruction, one can monitor the reconstruction error, defined as the mean squared error between the incoming features $p_t(x)$ and their reconstruction $\mathcal{R}(p_t(x_i))$.

$$r_e = \frac{1}{n} \sum_{i=1}^{n} (p_t(x_i) - \mathcal{R}(p_t(x_i)))^2 \tag{1}$$

We assume that when $\mathcal{R}$ is successful in reconstructing $p_t(x)$, by producing a sufficiently small reconstruction error r_e, it can be seen as an indicator that $\mathcal{R}$ well approximates $p_t(x)$ and that feature drift is successfully accounted for. On the other hand, in case $\mathcal{R}$ becomes unsuccessful in reconstructing $p_t(x)$, such that the reconstruction error r_e significantly deviates from zero, it can typically indicate a change to $p_t(x)$ and the underlying data generating process $\mathcal{P}(x)$.

Evidently, in order for $\mathcal{R}$ to successfully reconstruct incoming features, it has to be learned to capture the complexity and structure of the data. Capturing and modelling the internal structure of data is also critical for feature selection [11]. This implies the added benefit of exploiting the $\mathcal{R}$ component to rank the importance of the incoming features.

Therefore, in addition to drift detection, the FM approach can provide two important measures at a given time. Firstly, it can provide importance ranking of the original features under drift conditions. Secondly it also provides a fixed-size low dimension representation of $p_t(x)$ which can be exploited downstream.

3.2 Feature Selection Using Auto-Encoders

The FM approach is implemented using Auto-Encoders. It is conducted in an ensemble-like set-up consisting of multiple Auto-Encoders which are configured identically but receiving different pretraining histories. The general policy is to passively introduce different networks at different points into the data stream, as well as actively spawn new instances when a change is actively detected. The details about this are described in Sect. 3.3 below.

Feature selection is enabled through Sparse Learning in the encoding layer of the Auto-Encoder. To this end we integrate $||W||_{2,1}$ regularization in the first layer of the network to encourage row sparsity in the weight matrix [7]. For a matrix $M \in \mathbb{R}^{nxm}$, the $\ell_{2,1}$ norm is defined as:

$$||M||_{2,1} = \sum_{i=1}^{n} \sqrt{\sum_{j=1}^{m} m_{ij}^2} = \sum_{i=1}^{n} ||m^i||_2 \qquad (2)$$

This is achieved by taking the euclidean norm ($||m^i||_2 = \sqrt{m_{1i}^2 + ... + m_{ji}^2}$) of the columns of the weight matrix, and summing them for use as a penalty term in the loss function of the network to be minimized.

Combining the reconstruction loss in Eq. (1) with the sparse learning loss the objective function to be minimized is defined as:

$$\mathcal{L} = \frac{1}{n} \sum_{i=1}^{n} (p_t(x_i) - R(p_t(x_i)))^2 + \lambda \sum_{i=1}^{n} ||w^i||_2 \qquad (3)$$

where w^i is column i of the encoder weight matrix. The loss function is minimized by weight adjustment based on back-propagation of errors via a gradient-based stochastic optimization algorithm (ADAM) described here [9]. The feature ranking is done by inspection of weights. Given the described sparsity of the encoding weights, the ranking is obtained using:

$$f_{ranked} = diag(W_{(1)} W_{(1)}^T) \qquad (4)$$

3.3   Ensemble Policy

An ensemble-like set-up is employed to avoid that a single Auto-Encoder trained on one phase is unable to transit to the next phase in a timely manner. The purpose of the ensemble in this case is to process the stream in parts that are approximately stationary from the viewpoint of the Auto-Encoders. Note that only one instance from the ensemble is active at any time and generating feature rankings, since the reconstruction error r_e^i directly reveals the performance of each instance i. The ensemble policy can be seen as a *hybridized active and passive* policy, as it has elements of both active change detection based on the reconstruction error for a chunk and passive reinstantiation of unsuccessful instances at given intervals or number of chunks.

Ensemble policy: Inputs: n = number of samples between each passive reset of an old inactive instance, t relative error threshold value for triggering an actively detected change, r_e lowest average reconstruction error from the k previous samples based on a sliding window.

> Initialize: create first instance;
> **while** *incoming stream chunk c* **do**
> > **for** *each existing instance i* **do**
> > > train i on c;
> > > record instance chunk reconstruction error $\hat{r}_e^i$;
> >
> > **end**
> > **if** *lowest* $(\hat{r}_e^i - r_e)$ *is greater than t* **then**
> > > // Actively detected change;
> > > create new or reinstate oldest inactive instance;
> > > train new instance on c
> >
> > **end**
> > Update number of samples since last passive reset n_{last};
> > **if** n_{last} *greater than n* **then**
> > > // Passive reinstantiation;
> > > create new or reinstate oldest inactive instance;
> >
> > **end**
> > Update lowest sliding window average r_e based on last k samples
>
> **end**

Algorithm 1: Ensemble policy pseudocode

4   Results and Evaluation

The proposed online feature selection method has been tested on the public MNIST database to evaluate its efficacy. In the following we describe the test set-up in Sect. 4.1, the evaluation metrics in Sect. 4.2, and finally the results are presented in Sect. 4.3.

4.1   Test Set-Up

The input features used were the full 784 features of the training data from the MNIST public data set of handwritten digits. The test data stream was generated to consist of 3 phases of 25'000 instances: Odd, even and all digits respectively. Performance evaluation was done at the end of each phase using MNIST test data, not used in the data stream for learning. The test data consists of 10'000 handwritten digits not used by any of the data generators in creating the data streams.

4.2 Evaluation Metrics

The feature selection performance was evaluated by creating a Multi Layer Perceptron (MLP) classifier trained on generated training data from the data generator of the corresponding phase and tested on previously unseen test data. The feature selection performance evaluation used ground truth labels and was done offline.

The accuracy of the classifier is measured via the mean of the categorical accuracy over 10 iterations as defined by:

$$ACC(y, y_{predicted}) = \frac{1}{n_{samples}} \sum_{i=0}^{n_{samples}-1} 1(y_{predicted} = y) \tag{5}$$

4.3 Results

The results of testing an ensemble of 5 Auto-Encoders performing both change detection and feature selection are presented in Fig. 1 and Table 1 respectively. The results indicate that the feature selection performance remains stable throughout the 3 phases and with different data generation processes.

Table 1. Results of the feature selection performance at the end of each of the three phases.

Phase	1 (5cl, odd)	2 (5cl, even)	3 (5cl, odd)	3 (5cl, even)
100%	0.9818 ± 0.0013	0.9829 ± 0.0015	0.9856 ± 0.0009	0.9843 ± 0.0016
75%	0.9826 ± 0.0015	0.9841 ± 0.0015	0.9854 ± 0.0012	0.9859 ± 0.0018
50%	0.9837 ± 0.0010	0.9837 ± 0.0007	0.9844 ± 0.0015	0.9861 ± 0.0009
25%	0.9800 ± 0.0012	0.9812 ± 0.0011	0.9827 ± 0.0017	0.9826 ± 0.0023
10%	0.9526 ± 0.0033	0.9643 ± 0.0017	0.9566 ± 0.0033	0.9680 ± 0.0027
2%	0.8261 ± 0.0040	0.7730 ± 0.0030	0.8269 ± 0.0019	0.8679 ± 0.0023

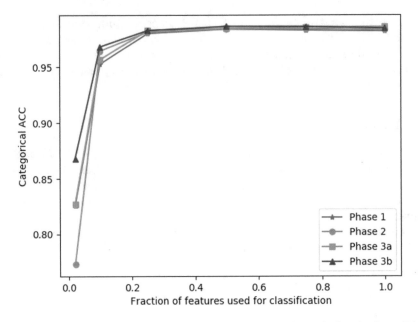

Fig. 1. Graph of the classification categorical accuracy at the end of each phase. Phase 3a, 3b evaluates the feature selection performance of odd and even digits respectively at the end of phase 3

5   Conclusions

A new method to online feature selection is proposed in this paper, which is implemented by feature monitoring using deep reconstruction networks as auto-encoders. An *active/passive* ensemble-like set of Auto-Encoders can successfully reconstruct data from a non-stationary data stream under feature drift conditions. This means that it can be advantageous to use reconstruction for the two combined purposes of detecting feature drift and also selecting and ranking incoming features. Allowing the downstream learning models, such as ensembles of classifiers to work with strictly reduced numbers of highly relevant features can provide significant benefits in terms of resource usage and learning performance.

References

1. Barddal, J.P., et al.: A survey on feature drift adaptation: definition, benchmark, challenges and future directions. J. Syst. Softw. **127**, 278–294 (2017). https://doi.org/10.1016/j.jss.2016.07.005. http://www.sciencedirect.com/science/article/pii/S0164121216301030. ISSN: 0164-1212
2. Calandra, R., et al.: Learning deep belief networks from non-stationary streams. In: Artificial Neural Networks and Machine Learning - ICANN 2012: 22nd International Conference on Artificial Neural Networks, Lausanne, Switzerland, 11–14 September 2012, Proceedings, Part II, Lecture Notes in Computer Science, vol. 7553, pp. 379–386. Springer, Heidelberg (2012). ISBN: 9783642332654

3. Ditzler, G., et al.: Learning in nonstationary environments: a survey. IEEE Comput. Intell. Mag. **10**(4), 12–25 (2015). https://doi.org/10.1109/MCI.2015.2471196. ISSN: 1556603X

4. Draelos, T.J., et al.: Neurogenesis deep learning. In: 2017 International Joint Conference on Neural Networks (IJCNN), pp. 526–533 (2017). https://doi.org/10.1109/IJCNN.2017.7965898

5. Han, K., et al.: Autoencoder inspired unsupervised feature selection. In: ICASSP, IEEE International Conference on Acoustics, Speech and Signal Processing - Proceedings 2018-April, pp. 2941–2945 (2018). https://doi.org/10.1109/ICASSP.2018.8462261. arXiv: 1710.08310. ISSN: 15206149

6. Hulten, G., Spencer, L., Domingos, P.: Mining time-changing data streams. In: Proceedings of the Seventh ACM SIGKDD International Conference on Knowledge Discovery and Data Mining, KDD 2001, pp. 97–106. ACM, San Francisco (2001). https://doi.org/10.1145/502512.502529. http://doi.acm.org/10.1145/502512.502529. ISBN: 1-58113-391-X

7. Ji, S., et al.: Feature selection based on structured sparsity: a comprehensive study. IEEE Trans. Neural Networks Learn. Syst. **28**(7), 1490–1507 (2016). https://doi.org/10.1109/tnnls.2016.2551724. ISSN: 2162-237X

8. Jolliffe, I.: Principal component analysis. In: Lovric, M. (ed.) International Encyclopedia of Statistical Science, pp. 1094–1096. Springer, Heidelberg (2011). https://doi.org/10.1007/978-3-642-04898-2_455. https://doi.org/10.1007/978-3-642-04898-2%7B%5C%7D455. ISBN: 978-3-642-04898-2

9. Kingma, D.P., Ba, J.: Adam: A Method for Stochastic Optimization, pp. 1–15 (2014). arXiv: 1412.6980. http://arxiv.org/abs/1412.6980

10. Li, J., Tang, J., Liu, H.: Reconstruction-based unsupervised feature selection: an embedded approach. In: IJCAI International Joint Conference on Artificial Intelligence, pp. 2159–2165 (2017). ISSN: 10450823

11. Li, J., et al.: Feature selection: a data perspective. ACM Comput. Surv. **50**(6), 94:1–94:45 (2017). https://doi.org/10.1145/3136625. http://doi.acm.org/10.1145/3136625. ISSN: 0360-0300

12. Nguyen, H.-L., et al.: Heterogeneous ensemble for feature drifts in data streams. In: Tan, P.-N., et al. (ed.) Advances in Knowledge Discovery and Data Mining, pp. 1–12. Springer, Heidelberg (2012). ISBN: 978-3-642-30220-6

13. Wang, S., Ding, Z., Fu, Y.: Feature selection guided auto-encoder. In: Thirty-First AAAI Conference on Artificial Intelligence (2017)

4-Rule Harmony Search Algorithm for Solving Computationally Expensive Optimization Test Problems

Ali Sadollah[1] ⓘ, Joong Hoon Kim[2(✉)] ⓘ, Young Hwan Choi[3] ⓘ, and Negar Karamoddin[4]

[1] Department of Mechanical Engineering,
University of Science and Culture, Tehran, Iran
[2] School of Civil, Environmental and Architectural Engineering,
Korea University, Seoul, South Korea
jaykim@korea.ac.kr
[3] Research Center for the Disaster and Science Technology, Korea University,
Seoul, South Korea
[4] Department of Chemistry, Semnan University, Semnan, Iran

Abstract. This paper proposes an enhanced harmony search algorithm for solving computationally expensive benchmarks widely used in the literature. We explored the potential and applicability of the original harmony search (HS) algorithm through introducing an extended version of the algorithm integrated with a new dynamic search equation enabling the algorithm to take guided larger steps at the beginning of the search. In the 4-Rule Harmony Search (4RHS), an extra rule is added to the standard HS without adding any user parameters to existing initial parameters. The 4RHS algorithm is then tested through optimal solving of different well-known and well-used benchmarks from classical to so CEC' 2015 series, where the results of the 4RHS are compared with simple and improved version of HS algorithms as well as other optimization techniques. The obtained optimization results show the attractiveness of the added rule into the standard HS.

Keywords: 4-Rule harmony search · Harmony search · Optimal design · Computationally expensive · Benchmark

1 Introduction

Optimization is widely used in engineering design problems, where the best values of design variables are determined with respect to one or more design performance criteria under a number of requirements and constraints. Different sources of complexity including uncertainty, nonlinearity, and dimensionality due to a large number of mixed continues and discrete variables and several discontinuous, non-differentiable, and even non-algebraic relationships among the variables, forming a nonconvex feasible space, can all make real-life optimization problems very challenging and difficult to solve by analytical, gradient-based solution techniques. Metaheuristic algorithms have, however, proved their capability to deal with these difficulties and to find near-optimal

© Springer Nature Switzerland AG 2020
J. H. Kim et al. (Eds.): ICHSA 2019, AISC 1063, pp. 202–209, 2020.
https://doi.org/10.1007/978-3-030-31967-0_23

solutions when analytical methods may fail, especially when the global optimum is surrounded by many local optima.

Harmony search (HS) algorithm, developed by Geem et al. [1], is derived from the concepts of musical improvisations and harmony knowledge. The HS algorithm and its improved variants have proved its advantages over other optimizers [2–7]. However, there is still room for improving the HS exploration capability as well as utilizing its potential in new, real-world design optimization applications.

This paper presents an extended version of the HS algorithm for optimal solving of classical and computationally expensive benchmarks by introducing an additional rule to three well-known rules of the standard HS algorithm. Then, we validate the performance of the proposed 4RHS through its application for solving well-known benchmark function optimization problems of different sizes and complexity.

The remainder of this paper is organized as follows. The next section describes the 4RHS algorithm in detail. Several well-known benchmarks such as classical and well-known series of CEC' 2015 benchmarks are taken into account for evaluation purposes of the performance and efficiency of the proposed method provided in Sect. 3, and conclusion and future research are provided in Sect. 4.

2 4-Rule Harmony Search

Recently, the HS has been applied to various research areas and obtained considerable attentions in different disciplines [8]. The HS intellectualizes the musical process of searching for a perfect state of harmony [1]. Musicians seek a fantastic harmony determined by aesthetic estimation. Similarly, optimization techniques seek a best state (global optimum) determined by an objective function value.

Aesthetic estimation depends on the set of the sounds played by a musical ensemble, whereas the objective function is evaluated by a set of adjustable variables. More aesthetic sounds can be produced by constant practice, and the optimization of the objective function can generally be improved by repeated iterations. Further details of HS can be found in the work of Geem et al. [1] and Kim et al. [2].

Standard HS consists of three general approaches. They are given as follows: (*i*) memory consideration phase, (*ii*) pitch adjusting phase, and (*iii*) uniform random search phase. The latter has been considered as exploration phase in HS similar to the concept of mutation in GAs. Accordingly, first and second phases (i.e., memory consideration and pitch adjusting) are related to the exploitation capability of HS. It is worth mentioning that memory consideration phase may act as exploration at early iterations due to the large diversity in the early populations.

However, looking at exploitation equations in standard HS and even IHS [3], we can see that they have suffered from having big guided jump at early iterations toward the best solution. As can be seen in particle swarm optimization (PSO) [9] movements toward the best experiences (i.e., global and personal experiences) lead us finding the optimum solutions. Therefore, in the proposed improved HS, we considered one additional rule for overcoming this drawback observed in standard HS and IHS. We have named this rule as "information exchange in HM". Let's see how it is related to the concept and terminology of original HS.

In standard HS, it is supposed that a musician randomly plays different notes and stores them in his/her memory (i.e., harmony memory). After that, based on a predefined probability (i.e., HMCR), he/she may extract some information from the memory or may not. After referring to the memory, he/she may alter some notes with respect to a probability (i.e., PAR). It is worth pointing out that at each try, he/she reminds the best harmonies. Therefore, after some efforts, he/she will come up with an enthusiastic harmony (i.e., considered as optimum solution).

As can be seen in original HS, there is no information exchange (i.e., experiences) among musicians. In pitch adjusting phase, a single musician adjusts a note (i.e., adding or subtracting) to fine tune his/her note. They do not consider or attract to the other musical players' notes. This exchange information (information flow) among musicians (instruments) may cause faster and more reliable outcomes.

Indeed, the concept of standard HS is based on the individual musician considering only his/her memory (note) or, better to say a musician acts individually. However, in the 4RHS, musical players utilize their experiences (among all best players) during the musical performance. It is obvious that in practical the outcome of such collaboration causes better performance.

One of the advantages of 4RHS compared with the standard HS is collaboration between exploration and exploitation phases caused by the suggested rule. New added rule, named as "information exchange in HM", performs more explorations rather than exploitation at early iterations due to large variation in quality solutions (i.e., harmonies) in HM. By iteration continuous, the added rule acts as exploitation operator by reason of small variation of solutions in HM.

In spite of adding one additional rule, unlike most of improved variants of HS, in this improved method, we did not add any user parameter to the original HS and IHS. Indeed, IHS [3] utilized the concept of using dynamic parameters instead of constant ones. They did not add any assumption and rules to the standard HS.

At first, based on the random HM, a new solution vector x^{New} is improvised. When improvising a new value for the ith decision variable x^{New} in the new vector, 4RHS algorithm utilizes one of the four operations (i.e., memory consideration, pitch adjusting, information exchange in HM, and random selection) as given follows:

$$x_i^{New} \leftarrow \begin{cases} x_i \in HM = \{x_i^1, x_i^2, \ldots, x_i^{HMS}\}, & w.p. \quad HMCR \\ x_i + BW \times (2rand - 1) \quad if \quad x_i \in HM, \quad or \quad x_i(k+m) \quad if \quad x_i(k) \in HM, & w.p. \quad HMCR \times PAR \\ x_i^{r(HM)} + 1.5 \times (2rand - 1) \times (x_i^{B(HM)} - x_i^{r(HM)}), & w.p. \quad HMCR \times (1 - PAR) \\ x \in \left[x_i^{Lower}, x_i^{Upper}\right] \quad or \quad \{x_i \in \{x_i(1), \ldots, x_i(k), \ldots, x_i(K)\}\}, & w.p. \quad (1 - HMCR) \end{cases}$$

$$(1)$$

where,

$w.p.$ means with respect to;

m is neighboring index which has normally 1 or -1 for discrete problems;

$\{x_i \in \{x_i(1), \ldots, x_i(k), \ldots, x_i(K)\}\}$ is the complete value set for the ith design variable;

$\{x_i^1, x_i^2, \ldots, x_i^{HMS}\}$ is the value set stored in HM for the ith decision variable;

$x_i^{B(HM)}$ is the best harmony in the HM for the ith design variable;

$x_i^{r(HM)}$ is a random harmony in the HM for the ith decision variable;

And *rand* is uniformly distributed random number between zero and one.

In the proposed version of HS, if the harmony is selected from HM and rejected by the pitch adjusting rule, then the new rule will be applied on the current harmony. First, a random harmony is selected from the HM (i.e., storage of best harmonies), afterwards, the chosen harmony moves to the best harmony in the HM. In fact, using the added rule, 4RHS exploits inside the harmony memory which contains the best harmonies. Therefore, in addition of pitch adjusting rule, this assumption (i.e., information exchange in HM) is considered as another exploration-exploitation phase for the HS.

As it is obvious, using dynamic parameters are preferred than constant parameters. Therefore, in the 4RHS, similar to IHS [3], probability value of PAR and bandwidth (BW) are changed adaptively at each iteration given as follows:

$$PAR(t) = PAR_{\min} + \frac{(PAR_{\max} - PAR_{\min})}{Max\_It} \times t \tag{2}$$

$$BW(t) = BW_{\max} \exp \left(\frac{Ln\left(\frac{BW_{\min}}{BW_{\max}}\right)}{Max\_It} \times t \right) \tag{3}$$

where PAR_{min} and PAR_{max} are the minimum and maximum pitch adjusting rates, BW_{min} and BW_{max} are the minimum and maximum bandwidths, t and Max_It are iteration counter and maximum number of iteration, respectively. Large PAR values with small BW values usually cause the improvement of best solutions in final iterations. With higher value for the PAR, the chance for the added rule (i.e., 4th rule) reduces. It means that, the new rule in this improved version of HS plays more important roles at early iterations when the probability value of PAR is small. Indeed, at early iterations, due to the high variance in HM, forcing harmonies to move to the best harmony can be assumed as guided exploitation inside the HM. By iteration goes, this chance will be reduced as the variation in HM decreases.

3   Validation and Efficiency

For validation of 4RHS, different types of benchmark problems including classical and CEC 2015 benchmarks series have been examined in this paper. Following sections validate performance and efficiency of the proposed improved method.

3.1   Classical Benchmark Problems

In order to examine the performance of 4RHS, several classical benchmark functions are selected and shown in Table 1. Reported classical benchmarks are classified into to the number of local minimum in unimodal (only one optimum) and multimodal (various optima), therefore Ackley, Griewank, Rastrigin, and Schwefel are multimodal functions, while Sphere and Schwefel 2.22 are unimodal functions.

Determined maximum number of function evaluations (NFEs) for the considered classical benchmarks was set to 50,000. The HS, IHS, global harmony search algorithm (GHS), self-adaptive global best harmony search algorithm (SGHS), intelligent GHS (IGHS), novel GHS (NGHS) have been considered for having comparison with the proposed improved HS [10]. The task of optimization for all reported optimizers was carried out for 30 independent runs.

Regarding the user parameters of the other reported variants of HS have been given in the literature [10]. After sensitivity analysis, the parameter values of the improved HS are given as follow: HMS of 30, HMCR of 0.98, PAR_{min} of 0.3, PAR_{max} of 0.99, BW_{min} of 0.0001, and BW_{max} of 0.01.

Table 2 tabulates statistical optimization results obtained by seven improved versions of HS for classical benchmark functions given in Table 1. Best obtained solution for each criterion has been highlighted in bold in Table 2. As can be seen in Table 2, except F1 and F4, the 4RHS has attained better statistical optimization results compared with the other considered improved versions.

3.2 CEC' 2015 Computationally Expensive Problems

In this section, the performance of 4RHS has been investigated using the well-known benchmarks in the literature so called CEC'15 computationally expensive optimization problems [11]. As reported in CEC 2015 (CEC'15), all test functions are minimization problems. Each function has a shift and rotated data which makes the optimization task difficult.

All test functions are shifted and scalable. For convenience, the same search ranges are defined for all test functions as $[-100,100]^D$. Table 3 shows properties of considered test problems used in the CEC'15. Detailed information regarding mathematical equations and shifted/rotated points has been given in the literature [11].

Based on the suggestion by the CEC'15, the optimization task has been carried out for 20 independent runs. Maximum NFEs of 500 and 1500 were considered as the stopping condition for dimension 10 and 30, respectively. Algorithms have been terminated when reaching the maximum NFEs or the error value $(F_i^* - F_i(x^*))$ is smaller than 0.001.

The optimization results obtained by the 4RHS are compared with standard HS (SHS), PSO, differential evolution (DE), real-coded genetic algorithm (RGA), $(\mu+\lambda)$-evolutionary strategy (ES), specialized and generalized parameters experiment of covariance matrix adaptation evolution strategy (CMAES-S and CMAES-G) [12] as given in Table 3. Note that the PSO, RGA, DE, SHS, and $(\mu+\lambda)$-ES in Table 3 have been coded and implemented in this research.

Table 1. Classical benchmark functions.

Function	Formulation	Interval
Sphere	$$\min F_1(X) = \sum_{i=1}^{D} x_i^2$$	$[-100, 100]^{D*}$
Ackley	$$\min F_2(X) = -20\exp\left(-0.2\sqrt{\frac{1}{D}\sum_{i=1}^{D} x_i^2}\right) - \exp\left(\frac{1}{D}\sum_{i=1}^{D}\cos(2\pi x_i)\right) + 20 + e$$	$[-32, 32]^D$
Griewank	$$\min F_3(X) = \frac{1}{4000}\sum_{i=1}^{D} x_i^2 - \prod_{i=1}^{D}\cos\left(\frac{x_i}{\sqrt{i}}\right) + 1$$	$[-100, 100]^D$
Schwefel 2.22	$$\min F_4 = \sum_{i=1}^{D}|x_i| + \prod_{i=1}^{D}|x_i|$$	$[-0.5, 0.5]^D$
Rastrigin	$$\min F_5(X) = \sum_{i=1}^{D}\left[x_i^2 - 10\cos(2\pi x_i) + 10\right]$$	$[-5.12, 5.12]^D$
Schwefel	$$\min F_6 = 418.9829 \times D - \sum_{i=1}^{D}(x_i\sin(\sqrt{|x_i|}))$$	$[-500, 500]^D$

*D is the problem dimension

Table 2. Statistical optimization results of seven improved HS for classical benchmark problems.

Func.	Measures	HS	IHS	GHS	SGHS	NGHS	IGHS	4RHS
F_1	Best	2.15E−07	2.43E+00	3.16E−01	1.35E−07	**9.89E−16**	1.80E−06	4.10E−07
	Worst	2.68E−07	3.10E+00	5.16E−01	4.31E−04	**4.94E−14**	6.18E−03	6.42E−07
	Average	2.17E−07	2.31E+00	3.47E−01	5.89E−02	**5.41E−15**	5.36E−04	5.18E−07
	SD	4.02E−08	8.62E−01	1.30E−01	4.69E−03	**9.96E−15**	1.40E−03	8.37E−08
F_2	Best	5.40E−01	4.30E−10	578.68E+00	5.85E−05	1.59E−08	6.90E−07	**4.10E−12**
	Worst	8.26E−01	4.50E−10	946.68E+00	7.44E−03	3.97E−07	3.92E−06	**5.49E−12**
	Average	5.83E−01	1.10E−10	609.72E+00	5.47E−04	4.43E−08	9.52E−07	**4.82E−12**
	SD	1.70E−01	2.20E−10	270.30E+00	1.49E−03	8.66E−08	8.99E−07	**3.58E−12**
F_3	Best	3.65E−02	4.05E−03	2.39E−05	2.33E−18	2.38E−04	1.10E−06	**2.01E−08**
	Worst	2.13E−01	1.04E−01	4.90E−05	3.45E−16	1.04E−01	3.53E−04	**3.41E−08**
	Average	5.38E−02	1.69E−02	2.57E−05	2.47E−17	7.15E−03	2.43E−05	**3.06E−08**
	SD	5.36E−02	2.86E−02	1.07E−05	6.51E−17	2.63E−02	6.98E−05	**1.34E−08**
F_4	Best	4.30E−01	7.30E−02	4.00E−04	**1.33E−16**	2.64E+00	5.85E−05	2.12E−03
	Worst	6.91E−01	2.77E−01	2.71E−03	**6.31E−15**	2.88E+00	7.44E−03	2.85E−03
	Average	4.23E−01	9.40E−02	4.73E−04	**5.42E−16**	2.41E+00	5.47E−04	2.56E−03
	SD	1.14E−01	6.60E−02	4.70E−04	**1.20E−15**	8.19E−01	1.49E−03	2.57E−04
F_5	Best	3.48E−04	6.90E−01	1.69E−01	9.99E−02	2.90E−10	1.54E−10	**6.47E−12**
	Worst	3.60E−04	8.78E−01	2.18E−01	9.80E−02	4.79E−10	3.66E−10	**1.01E−11**
	Average	3.40E−04	7.05E−01	1.72E−01	3.29E+00	2.77E−10	1.97E−10	**8.92E−11**
	SD	2.69E−05	1.35E−01	3.70E−02	8.54E−01	3.96E−10	1.15E−10	**1.22E−10**
F_6	Best	7.61E−02	7.26E−02	1.84E−02	3.25E−02	1.20E−02	6.42E−04	**3.81E−04**
	Worst	1.80E−01	1.55E−01	4.08E−02	6.01E−02	5.69E−02	3.15E−02	**3.81E−04**
	Average	8.60E−02	7.80E−02	1.89E−02	3.44E−02	1.98E−02	2.02E−03	**3.81E−04**
	SD	3.71E−02	2.79E−02	7.25E−03	1.01E−02	1.63E−02	5.78E−03	**7.07E−09**

Table 3. Comparison of average error values for different optimizers for F_1 to F_{15}.

Fun.	D	PSO	DE	RGA	$(\mu+\lambda)$-ES	SHS	CMAES-S	CMAES-G	4RHS
F_1	10	1.21E+09	3.41E+09	1.00E+09	2.63E+09	7.68E+08	**3.66E+07**	6.59E+07	2.47E+08
	30	3.69E+09	2.39E+10	2.28E+10	3.57E+10	2.30E+09	**6.87E+07**	1.10E+08	1.05E+09
F_2	10	3.33E+04	7.49E+04	**2.86E+04**	4.84E+04	7.21E+04	5.80E+04	1.02E+05	4.24E+04
	30	**7.04E+04**	1.82E+05	7.90E+04	1.61E+05	1.35E+05	2.36E+05	2.95E+05	1.23E+05
F_3	10	3.10E+02	3.10E+02	3.09E+02	3.10E+02	3.08E+02	6.12E+02	6.15E+02	**3.06E+02**
	30	3.33E+02	3.41E+02	3.31E+02	3.43E+02	**3.28E+02**	6.33E+02	6.52E+02	3.29E+02
F_4	10	2.15E+03	2.29E+03	1.96E+03	1.43E+03	7.86E+02	3.18E+03	4.10E+03	**7.24E+02**
	30	6.12E+03	7.96E+03	6.07E+03	7.05E+03	1.57E+03	8.67E+03	1.20E+04	**1.48E+03**
F_5	10	5.02E+02	5.02E+02	5.02E+02	5.03E+02	5.02E+02	1.00E+03	1.00E+03	**5.02E+02**
	30	5.03E+02	5.04E+02	**5.02E+02**	5.04E+02	5.04E+02	1.00E+03	1.00E+03	5.04E+02
F_6	10	6.01E+02	6.02E+02	6.02E+02	6.02E+02	6.01E+02	1.20E+03	1.20E+03	**6.00E+02**
	30	6.00E+02	6.03E+02	6.03E+02	6.04E+02	6.00E+02	1.20E+03	1.20E+03	**6.00E+02**
F_7	10	7.07E+02	7.25E+02	7.11E+02	7.16E+02	7.05E+02	1.40E+03	1.40E+03	**7.03E+02**
	30	7.02E+02	7.54E+02	7.39E+02	7.82E+02	7.03E+02	1.40E+03	1.40E+03	**7.00E+02**
F_8	10	9.99E+02	4.16E+03	1.04E+03	1.16E+03	8.61E+02	1.61E+03	1.64E+03	**8.22E+02**
	30	1.03E+04	7.99E+05	2.79E+05	7.37E+06	8.47E+03	**1.76E+03**	2.32E+03	2.59E+04
F_9	10	9.03E+02	9.04E+02	9.04E+02	9.04E+02	9.04E+02	1.80E+03	1.80E+03	**9.03E+02**
	30	9.13E+02	9.13E+02	9.13E+02	9.14E+02	9.13E+02	1.82E+03	1.82E+03	**9.13E+02**
F_{10}	10	4.71E+05	1.36E+06	3.47E+05	1.46E+06	1.33E+06	**1.74E+05**	1.77E+06	3.53E+05
	30	3.65E+06	3.87E+07	1.70E+07	9.53E+07	3.69E+07	**3.63E+06**	1.47E+07	1.43E+07
F_{11}	1	1.11E+03	1.12E+03	1.11E+03	1.11E+03	1.11E+03	2.21E+03	2.21E+03	**1.10E+03**
	30	1.16E+03	1.28E+03	1.21E+03	1.43E+03	**1.14E+03**	2.24E+03	2.25E+03	1.18E+03
F_{12}	10	1.50E+03	1.59E+03	1.46E+03	1.57E+03	1.55E+03	2.73E+03	2.98E+03	**1.43E+03**
	30	1.97E+03	3.01E+03	**1.72E+03**	3.8E+03	2.79E+03	3.45E+03	4.09E+03	1.84E+03
F_{13}	10	1.66E+03	1.79E+03	1.65E+03	1.66E+03	1.64E+03	3.25E+03	3.30E+03	**1.64E+03**
	30	1.72E+03	1.96E+03	1.93E+03	2.22E+03	1.72E+03	3.38E+03	3.42E+03	**1.70E+03**
F_{14}	10	1.61E+03	1.61E+03	1.60E+03	1.61E+03	1.61E+03	3.20E+03	3.21E+03	**1.60E+03**
	30	**1.67E+03**	1.74E+03	1.69E+03	1.84E+03	1.70E+03	3.26E+03	3.30E+03	1.70E+03
F_{15}	10	1.89E+03	1.94E+03	**1.89E+03**	1.97E+03	2.05E+03	3.77E+03	3.90E+03	1.95E+03
	30	2.65E+03	2.93E+03	**2.68E+03**	2.91E+03	2.46E+03	4.42E+03	4.83E+03	2.60E+03

4   Conclusions and Future Research

In this paper, an improved version of harmony search (HS) algorithm is proposed. Aiming to improve the exploration capabilities of HS, a new rule has been added to the three well-known rules of standard HS so called 4-Rule harmony search (4RHS). A new dynamic search equation enabling the HS algorithm to take guided larger steps at the beginning of the search without adding new user parameters is considered as improvement to the standard HS. Afterwards, several classical and well-known benchmarks so called CEC' 2015 have been examined to show the performance and efficiency of the proposed HS over other variants of HS as well as other existing optimizers. Obtained optimization results prove the superiority of 4RHS compared with the reported optimizers in terms of statistical optimization results. As further research, considering constrained and multi-objective optimization problems can be beneficial to see the performance of 4RHS over wide range (nature) of optimization problems. Besides, real-life applications in engineering for instance optimal sizing of water distribution networks having many design variables are the prime interest of our future studies.

Acknowledgement. This work was supported by the National Research Foundation (NRF) of Korea under a grant funded by the Korean government (MSIP) (NRF-2019R1A2B5B03069810).

References

1. Geem, G.W., Kim, J.H., Loganathan, G.V.: A new heuristic optimization algorithm: harmony search. Simulation **76**(2), 60–68 (2001)
2. Kim, J.H., Geem, Z.W., Kim, E.: Parameter estimation of the nonlinear Muskingum model using harmony search. J. Am. Water Resour. Assoc. **37**(5), 1131–1138 (2001)
3. Mahdavi, M., Fesanghary, M., Damangir, E.: An improved harmony search algorithm for solving optimization problems. Appl. Math. Comput. **188**(2), 1567–1579 (2007)
4. Geem, Z.W., Sim, K.B.: Parameter-setting-free harmony search algorithm. Appl. Math. Comp. **217**(8), 3881–3889 (2010)
5. Gao, X.Z., Wang, X., Ovaska, S.J.: Uni-modal and multi-modal optimization using modified harmony search methods. Int. J. Innov. Comput. Inf. Control **5**(10A), 2985–2996 (2009)
6. Manjarres, D., Landa-Torres, I., Gil-Lopez, S., DelSer, J., Bilbao, M.N., Salcedo-Sanz, S., Geem, Z.W.: A survey on applications of the harmony search algorithm. Eng. Appl. Artif. Intell. **26**, 1818–1831 (2013)
7. Fesanghary, M., Damangir, E., Soleimani, I.: Design optimization of shell and tube heat exchangers using global sensitivity analysis and harmony search algorithm. Appl. Therm. Eng. **29**(5–6), 1026–1031 (2009)
8. Yoo, D.G., Kim, J.H., Geem, Z.W.: Overview of harmony search algorithm and its applications in civil engineering. Evol. Intell. **7**, 3–16 (2014)
9. Kennedy, J., Eberhart, R.C.: Particle swarm optimization. In: IEEE International Conference on Neural Networks, Piscataway, NJ, pp. 1942–1948 (1995)
10. Zhang, P., Ouyang, H., Gao, L.: Improved harmony search algorithm with perturbation strategy. In: 27th Chinese Control and Decision Conference (CCDC) (2015). https://doi.org/10.1109/CCDC.2015.7162873
11. Chen, Q., Liu, B., Zhang, Q., Liang, J.J., Suganthan, P.N., Qu, B.Y.: Problem definition and evaluation criteria for CEC 2015 special session and competition on bound constrained single-objective computationally expensive numerical optimization. Computational Intelligence Laboratory, Zhengzhou University, Zhengzhou China and Nanyang Technological University, Singapore, Technical report (2014)
12. Andersson, M., Bandaru, S., Ng, A., Syberfeldt, A.: Parameter tuned CMA-ES on the CEC'15 expensive problems. In: IEEE CEC Evolutionary Computation, pp. 1950–1957 (2015)

An Empirical Case Study for the Application of Agricultural Product Traceability System

Yu Chuan Liu[1(⊠)] and Hong Mei Gao[2]

[1] Tainan University of Technology, Tainan 71002, Taiwan
t00258@mail.tut.edu.tw
[2] Tianjin Agriculture University, Tianjin 300384, China

Abstract. Globalization of the food supply chain has exacerbated the concerns for food-related safety issues and driven the demand for more information about the vertical and horizontal food supply chain. Transparency in the way agriculture was grown and handled throughout the supply chain process resulted in an important issue of 'traceability' in global food trade and safety. Traceability is an essential subsystem of quality management and must be managed by setting up a traceability system, which keeps the data tracking of product routes and of selected attributes. A collaborative research project for the supply and logistics of Tianjin Shawo radish was proposed recently. An end-to-end mobile application that traces the farming activities by using mobile handheld devices to capture information of farming operations is developed and implemented. Requirements analysis and system development for the traceability information of the food supply chain is studied. Applications of this mobile contract farming information management system show that the complicated manual traceability data recording can be significantly reduced for the farmers. The consumers' confidence for healthy food choices with clear food traceability can be improved.

Keywords: Food traceability · Logistics information management system · Contract farming

1 Introduction

Issues for the safety of food planting, processing, the environmental and ecological impact of agriculture had been exacerbated by the incidences such as the human form of bovine spongiform encephalitis (BSE, mad cow disease), genetically engineered foods, and contamination of fresh and processed agriculture. The heightened awareness of food-related safety issues among food consumers drives the demand for more transparency about the vertical food supply chain. A traceability system can consist two elements, the routes of the product, path along which products can be identified throughout the manufacturing, distribution and retail procedures, and the extent of traceability wanted [1]. Food traceability system is aimed to provide information visibility through the farming, production, packing, distribution, transportation, and sales process, and receiving enthusiastic research interests due to the food supply chain globalization. Food traceability requires that all stakeholders within the food supply chain, including agriculture and feed producers, food manufacturers, retailers, etc.,

© Springer Nature Switzerland AG 2020
J. H. Kim et al. (Eds.): ICHSA 2019, AISC 1063, pp. 210–217, 2020.
https://doi.org/10.1007/978-3-030-31967-0_24

must be able to identify the source of all raw materials and ingredients and also to whom the products have been sold. The food companies must apply identification systems and data handling procedures and these must be integrated into their quality management system. The sector encompassing information technology (IT) to find a reasonable compromise between the simple, step by step passing of traceable unit IDs for the neighboring actors, and the accumulated enormously huge databases of the actors. The traceability system is to provide services for the supply chain actors on cooperative basis of the mutual interests [2].

Opara [3] reviewed the concepts of supply chain management and traceability in agriculture and highlighted the technological challenges, including food product label and identification, activity/process characterization, information systems for data capture, analysis, storage and communication, and the integration of the overall traceable supply chain in implementing traceable agricultural supply chains. Wang et al. [4] addressed that the values on traceability can be integrated with the supply chain management processes to manage the business process and improve its performance. Although a recall action could be absolutely critical for a company, both in terms of incurred costs and of media impact, at present most companies do not posses reliable methods to precisely estimate the amount of product that would be discarded in the case of recall. Considering the food traceability system as part of the logistics management, Bosona and Gebresenbet [5] summarized the literature review on the food traceability issues. The definition, driving forces, barriers, benefits, traceability technologies, improvements, and performances of food traceability system had been discussed. It was pointed out that the development of full chain food traceability system is quite complex in nature, and a deeper understanding of real processes from different perspectives such as economic, legal, technological, and social issues are essential.

The major IT development lines, the support potential of their integration, organizational requirements for the utilization, and possible consequences for the future organization of the agro-food sector were reviewed [6]. A new model and prototype of a new Farm Information Management System, which meets the changing requirements for advising managers with formal instructions, recommended guidelines and documentation requirements for various decision making processes, was developed [7]. As achieving end-to-end traceability across the supply chain is quite a challenge from a technical, a co-ordination and a cost perspective, Kelepouris et al. [8] suggested a radio frequency identification (RFID) technology and outlined both information data model and system architecture that made traceability feasible and easily deployable across a supply chain. Based on an integration of alphanumerical codes and RFID technology, the traceability system for Parmigiano Reggiano (the famous Italian cheese) was developed [9]. Manthou et al. [10] provided empirical insights regarding the use of Internet-based applications in the agri-food supply chain by focusing on the Greek fruit canning sector. A PDA-based Record-keeping and Decision-support System for traceability in cucumber production was developed on Windows Mobile platform invoking a Geographic Information System (GIS) control [11]. The state-of-the-art review in the recent advancements of food processing and packaging industry in the fields of smart packaging and materials, automation and control technology, standards, and their application scenarios, and production management principles and their improvements was proposed [12].

Liu and Gao [13] applied the model-driven business transformation for the review and analysis of the logistics management of fresh agricultural products supply chain. The purpose of this research is to further provide an empirical case study for the traceability of the famous Shawo radish in Tianjin. The mobile contract farming information system was implemented. Solution for the food safety supply chain will be crucial for the integration of information and logistics management in fresh agricultural products supply chain business.

2   Mobile Solution from Model-Driven Business Transformation

A full life-cycle business-to-technology method, model-driven business transformation MDBT [14], is both a business transformation methodology and a set of innovative technologies that allow business strategies to be realized by choreographing workflow tools and human activities. MDBT uses formal models to explicitly define the structure and behaviour of a business component. The framework is made up of four layers: business strategy, business operations, solution composition, and IT implementation. Each layer constitutes a different level of abstraction, performs a well-defined function, and has a different audience. In the MDBT approach, the transformation process begins with the identification of the strategic goals and objectives of the business component. This leads to a set of initiatives that support these goals. These initiatives determine the definition, analysis, optimization, and implementation of the business operations of the organization such that the strategic goals can be achieved. Formal definition of the business operations and the operational KPIs (key performance indicators) is the next step of transformation process which was referred as the business operation model. A business operation model defines the key business artefacts and the operations performed on these artefacts. The third step of solution composition in MDBT is the judicious use of technology to support the execution of business operations. This involves the generation of a platform-independent solution composition model and the realization of this model on a specific software platform. The final step in MDBT is to create an implementation of the IT solution on a specific IT platform.

Figure 1 shows that the logistics information flow between the supply chain enterprises can be shared and integrated for further enhancement of the supply chain management efficiency. The logistics information during the transportation and storage for all the supply chain stages to ensure the completeness of the traceability management. The information system must be designed to cope with the intricate farming data, food processing information, and the rigmarole product transportation and storage information. As the authors reviewed the current fresh agricultural products supply chain in Tianjin, the first two issues of the are the product loss during the logistics procedure and cost (and/or effectiveness) of the cold chain logistics. The objectives are consequently to improve the effectiveness and efficiency for fresh agricultural products logistics management. The KPIs for hardware protection can be fresh agricultural products loss during transportation, handling, and storage procedures. The KPIs for the software management can be the easy and complete access of the traceability data through the whole supply chain process, i.e., the farming data, process and ingredient information, the transportation and storage environment condition histories.

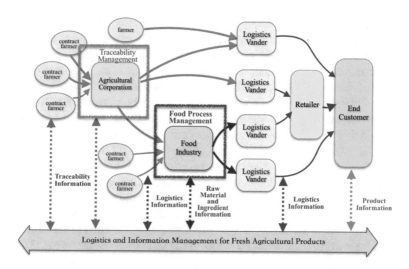

Fig. 1. Infrastructure of the logistic management for agricultural products.

System architecture for the traceability information management of agricultural corporations is shown in Fig. 2. The farming data are collected by the mobile devices that can provide data communication anytime and anywhere through the mobile 4G or Wi-Fi network. The data to be collected for traceability including the farming activities messages such as seeding, weeding, fertilizing, disease prevention, harvest and shipping, are designed to be stored in Quick response codes, QR codes. By scamming the QR code with mobile device, the data stored in the QR code can be decoded and enabling a transaction to be uploaded to the traceability database [13].

Fig. 2. System requirements of the farming traceability information management.

Operation scenarios for farming activity message collection are analyzed. The farming information collection is started by scanning the corresponding QR code label. The mobile device will decode and link to the application server and enabling the farmer authentication transaction. The farmer should input the proper username and password and the mobile system will verify the validation. As the verification process completed, the decoded farming operation message will be shown and start the confirm request transaction. The operation messages will be uploaded to the database after the confirm signal is entered. For the case that more detail attributes are required to be recorded, the detail activity messages request transaction will be enabled after finishing the authentication. Detailed operation scenario for the pesticide usage recording process can be found in [15]. By properly encoding the farming messages to the QR codes, the messages can be read by the mobile devices and uploaded simultaneously to the application server of the mobile farming system. The farming activity information can be captured by simply scanning the QR code labels and uploaded in real time, instead of manual handwriting and input by batch to the traceability system (Fig. 3).

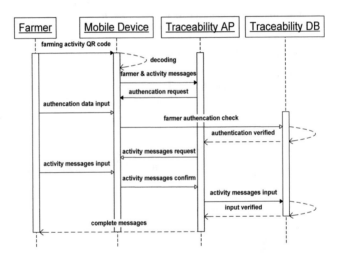

Fig. 3. Sequential diagram for the operation scenario of real-time farming activity message upload.

3   Traceability System for Gin-Guo Pear in Tianjin

Traceability system in China is attracting more and more research and industrial interests. The academic cooperation across between Tainan University of Technology in Taiwan and Tianjin University of Agriculture in Tianjin has been started since 2013. Researches about the food traceability in TUA have got the financial support from the Spark Program 2013–2015 of the Ministry of Science and Technology of China. The Auodong village is one of the precise poverty alleviation villages in the Wuqing district of Tianjin. The Gin-Guo pear planting was introduced since 2016 to enhance the economic revenue. They formed Auodong corporate and joined the TUA project in

2018 to implement the mobile farming information system to build their own traceability system. The aim of the traceability system is not only to improve the safety traceability but also to promote the value of their brand, Yunbei Gin-Guo Pear, among the others of Gin-Guo pear farms. The implementation of mobile farming data collection of the pilot project can be summarized as following.

(i) Kickoff meeting
     The pilot project began by the kickoff meeting hold by Auodong cooperative, government employees, and the technical team form Tainan University of Technology and Tianjin University of Agriculture.
(ii) Farming data and QR code definition
     The system analysis for the production activities and related information required for Gin-Guo pear farming was firstly reviewed, including farming activities, fertilizer, and pesticides, were applied for the pilot system. The QR code for every farming activities were encoded after the data definition decided.
(iii) Database and brief traceability system development
     The farming activities, such as irrigation, weeding, branch trimming, etc., are all recorded by scanning the corresponding QR code label as shown in Fig. 4.

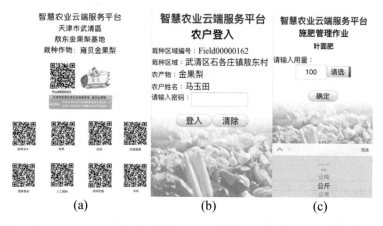

(a)   (b)   (c)

Fig. 4. Operation scenario for the pesticide usage recording process, (a) scanning the QR code, (b) user authentication, (c) the message input transaction enabled.

(iv) Mobile farming SOP and training
     To provide effective training for the mobile farming information system, standard operation procedure was established including the brief operation instructions of the hardware device, the QR code scanning process, and the usual questions and answers for the usage.
(v) Field testing and auditing
     The field test for the mobile communication at the farmland is necessary to guarantee the effectiveness of the system. The field testing and auditing were performed at the site of farmland in Tianjin. The data communication through mobile 4G networking is worked effectively at the farmland.

The home page for the back-end server of the mobile platform is shown in Fig. 5a. As a private traceability service platform, the home page of the service platform shows the organizations registered in this sector. After logged in to the system by the administrator of the organization, the production data of farm information, farmer, product, fertilizer, pesticides, harvesting, and packing information can all be maintained at the platform. The farm information (Fig. 5b) includes: name of the contact farmer, telephone and fax numbers, e-mail, address, facebook, and the multimedia video is also provided in accordance with the text information and pictures. The farming data uploaded can be queried and only the administrator can delete or edit the farming data to ensure the correctness and reliability (Fig. 5c). By scanning the traceability label, the brief traceability information, including name of the organization and producer, traceability ID number, telephone number, and location of the farmland, is displayed on the mobile device (as shown in Fig. 5b). The detail farm and traceability information can be obtained by further clicking the corresponding linkage.

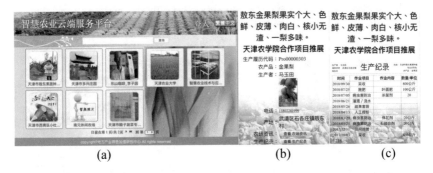

(a)   (b)   (c)

Fig. 5. Mobile server platform user interfaces, (a) the home page, (b) farmer information, (c) traceability data.

4   Summary and Conclusions

Implementation of traceability system for agriculture and food supply chains is necessary to enhance the competence of product. Traceability data involves miscellaneous records that mostly kept by manual handwriting. To reduce the minute and unreliable traceability data recording procedure, the mobile farming information system is developed to enhance the efficiency of e-traceability. The operation scenarios for the mobile traceability data construction and collection are studied, the architecture for the system and database are schemed, and the application examples are proposed in this research. By properly encoding the farming activity messages to the QR code labels, the farming data can be read by the mobile devices and uploaded simultaneously to the mobile platform server. The farming activity information can be captured by simply scanning the QR code labels and uploaded in real time. The miscellaneous traceability data collection can be significantly reduced. The results showed that the food traceability can be more credible due to the farming data is collected more reliable.

Acknowledgments. Authors wishing to acknowledge financial support from Tianjin Administration of Foreign Affairs (project no. JXB2019017) and the Wuqing Science and Technology Development Council (project no. WQKJ201803).

References

1. Moe, T.: Perspectives on traceability in food manufacture. Trends Food Sci. Tech. **9**, 211–214 (1998)
2. Varga, M., Csukas, B.: On the way toward the sector spanning agrifood process traceability. Agric. Inform. **1**(1), 8–18 (2010)
3. Opara, L.U.: Traceability in agriculture and food supply chain: a review of basic concepts, technological implications, and future prospects. J. Food Agric. Environ. **1**(1), 101–106 (2003)
4. Wang, X., Li, D., Li, L.: Adding value of food traceability to the business: a supply chain management approach. Int. J. Serv. Oper. Inform. **4**(3), 232–257 (2009)
5. Bosona, T., Gebresenbet, G.: Food traceability as an integral part of logistics management in food and agricultural supply chain. Food Control **33**, 32–48 (2013)
6. Schiefer, G.: New technologies and their impact on the agri-food sector: an economists view. Comput. Electron. Agric. **43**(2), 163–172 (2004)
7. Sørensena, C.G., Fountasb, S., Nashf, E., Pesonend, L., Bochtisa, D., Pedersene, S.M., Bassoc, B., Blackmoreg, S.B.: Conceptual model of a future farm management information system. Comput. Electron. Agric. **72**(1), 37–47 (2010)
8. Kelepouris, T., Pramatari, K., Doukidis, G.: RFID-enabled traceability in the food supply chain. Ind. Manag. Data Syst. **10**(2), 183–200 (2007)
9. Regattieri, A., Gamberi, M., Manzini, R.: Traceability of food products: general framework and experimental evidence. J. Food Eng. **81**(2), 347–356 (2007)
10. Manthou, V., Matopoulos, A., Vlachopoulou, M.: Internet-based applications in the agrifood supply chain: a survey on the Greek canning sector. J. Food Eng. **70**(3), 447–454 (2005)
11. Li, M., Qian, J.P., Yang, X.T., Sun, C.H., Ji, Z.T.: A PDA-based record-keeping and decision-support system for traceability in cucumber production. Comput. Electron. Agric. **70**(1), 69–77 (2010)
12. Mahalik, N.P., Nambiar, A.N.: Trends in food packaging and manufacturing systems and technology. Trends Food Sci. Tech. **21**(3), 117–128 (2010)
13. Liu, Y.C., Gao, H.M.: System analysis for the traceability and logistics management of fresh agricultural products supply chain. Adv. Econ. Bus. Manag. Res. **50**, 100–104 (2017)
14. Kumaran, S.: Model driven enterprise. In: Proceedings of the Global Enterprise Application Integration Summit, Banf, Canada, pp. 166–180 (2004)
15. Liu, Y.C., Gao, H.M.: Development and applications of mobile farming information system for food traceability. Health Manag. Appl. Comput. Med. Health **12**, 244–268 (2015)

Study on the Coordinating Policy of Multi-echelon Inventory System

Tao Yang[✉]

Shanghai Polytechnic University, Shanghai 201209, China
yangtao@sspu.edu.cn

Abstract. The mechanism to enhance the performance of the supply chain by coordination policy is analyzed in detail. The simple supply chain is modeled as a decentralized two-level inventory system, and the optimal decision formulas are deduced for the firms at their stages in the supply chain by systematic optimality. The result of numerical experiment shows that the coordination failure is observed in the simple deterministic decentralized inventory system. It is proved obviously that the proper revenue sharing contract design is a useful coordinating tool to incentive the firms to choose actions which lead firms and total supply chain to global optimization as a result.

Keywords: Coordinating policy · Global optimization · Decentralized inventory system

1 Introduction

The well known economic ordering quantity model was firstly developed to solve the optimal inventory control problem for a isolated inventory system [1]. Recently, more and more researches were involved in decision problem of the multi-echelon inventory system and large amount of achievements about the supply chain management have also emerged [2–4]. Normally, centralized inventory system leads to higher performance than decentralized inventory system [5]. Actually, the coordination policy, such as revenue sharing contract, has became an useful method to improve the total supply chain performance [6,7].

This paper differs from other undeterminate decentralized inventory system models [8,9], focuses on the deterministic decentralized multi-echelon inventory system, and then develops new coordinating policy. Next section discusses the optimal decision of decentralized inventory system.

2 Optimal Decision of Decentralized Inventory System

2.1 The Structure of Decentralized Two-Level Inventory System

We consider a two-level inventory system of one-supplier and one-retailer as in Fig. 1. The upstream supplier is the node 1, and the downstream retailer is the

© Springer Nature Switzerland AG 2020
J. H. Kim et al. (Eds.): ICHSA 2019, AISC 1063, pp. 218–225, 2020.
https://doi.org/10.1007/978-3-030-31967-0_25

node 2. The supplier products some goods and sells them to the retailer. Then the retailer sells the goods to its customer on market price.

The order of decision making is: firstly the supplier selects a wholesale price, then the retailer chooses an order quantity, and lastly the supplier determines its batch production.

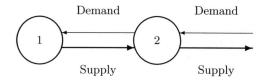

Fig. 1. The structure of decentralized two-level inventory system

For convenience, the notations of the paper are summarized below:

c_i to represent the unit production cost if $i = 1$;
    to represent the wholesale price if $i = 2$.
p the retail price at final market.
Q_i the order quantity for node i, $i = 1, 2$.
K_i the setup cost at node i, $i = 1, 2$.
D the demand per unit time.
I the holding cost coefficient per unit time.
π_i the long-term average profit for node i, $i = 1, 2$.
α the revenue share of the supplier.
n Q_1 is an integer n times to Q_2.
T_i the order cycle time at node i, $i = 1, 2$.
π_D the total long-term average profit of a decentralized system
π_C the total long-term average profit of a centralized system

2.2 Decision for Node 2

For node 2, its decision problem is the classic economic ordering quantity model. Suppose the retailer encounter a deterministic demand D without lead time, as Fig. 1. The objective of node 2 is to maximize its long-term average profit by an optimal order quantity.

Based on the cost analysis as in [1,2], the long-term average profit function for node 2 is:

$$\pi_2(Q_2) = D(p - c_2) - \frac{1}{2}Q_2 I c_2 - \frac{K_2 D}{Q_2} \tag{1}$$

The objective of node 2 is maximize its profit π_2. The π_2 is a real function of Q_2, By the first order necessarily condition of optimality of function $\pi_2(Q_2)$, we conclude the optimal decision of the node 2 as follows.

$$Q_2^* = \sqrt{\frac{2DK_2}{Ic_2}} \tag{2}$$

$$T_2^* = \frac{Q_2^*}{D} = \sqrt{\frac{2K_2}{DIc_2}} \tag{3}$$

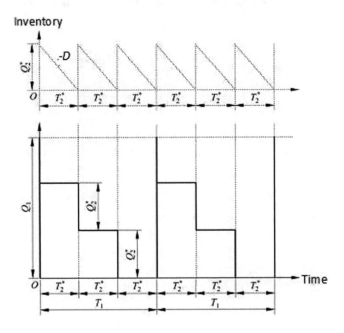

Fig. 2. The decision model of two-level inventory system

2.3   Decision for Node 1

For node 1, its decision problem is not the classic economic ordering quantity model. The demand of node 1 is the optimal ordering quantity of the node 2, as shown in Fig. 1.

From the research result of [3] and [4], the optimal production batch of the supplier must be an integer n times to the order quantity of the retailer, as shown in Fig. 2. So we have Eq. 4

$$Q_1 = nQ_2^* \tag{4}$$

Therefore, the objective of node 1 is to maximize its long-term average profit by an optimal integer multiplier n. The long-term average profit function of node 1 can be concluded:

$$\pi_1(n) = D(c_2 - c_1) - \frac{K_1 D}{nQ_2^*} - \frac{n-1}{2} Q_2^* I c_1 \tag{5}$$

Now we take n as a real number, so the π_1 is a real function of n. By first order necessarily condition of optimality of $\pi_1(n)$, the optimal solution is as Eq. 6

$$n = \sqrt{\frac{K_1 c_2}{K_2 c_1}} \tag{6}$$

By the round rule

$$n^* = \begin{cases} \lceil n \rceil, \text{ if } \pi_1(\lceil n \rceil) > \pi_1(\lfloor n \rfloor); \\ \lfloor n \rfloor, \text{ otherwise.} \end{cases} \tag{7}$$

We take an approximation integer of n as the optimal decision of the node 1.

$$Q_1^* = n^* Q_2^* \tag{8}$$

$$\pi_D(n, Q_2) = \pi_1(n) + \pi_2(Q_2) \tag{9}$$

3   Coordination Failure of Decentralized Inventory System

A decentralized inventory system is defined as coordination failure when its performance is less than the global optimal result of a same centralized inventory system. Even if firms choose individually optimal actions [6], the coordination failure always happen because each firm makes decisions based on their own margin, not the supply chain's margin, which is also called double marginalization [7].

In a centralized inventory system, as in Fig. 3, the node 2 makes decision of an optimal order quantity Q_2, and the node 1 makes decision of an optimal integer multiplier n simultaneously. The profit function of the node 2 is the same as Eq. 1. The other profit functions is described as follows.

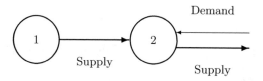

Fig. 3. The structure of centralized two-level inventory system

$$\pi_1(n, Q_2) = D(c_2 - c_1) - \frac{K_1 D}{n Q_2} - \frac{n-1}{2} Q_2 I c_1 \tag{10}$$

$$\pi_C(n, Q_2) = \pi_1(n, Q_2) + \pi_2(Q_2) \tag{11}$$

By the K-K-T condition of optimality of function π_C

$$n = \sqrt{\frac{K_1(c_2 - c_1)}{K_2 c_1}} \tag{12}$$

$$Q_2 = \sqrt{\frac{2D(\frac{K_1}{n} + K_2)}{I(nc_1 - c_2 - c_1)}} \tag{13}$$

$$\pi_C(n) = Dp - \sqrt{2DI(\frac{K_1}{n} + K_2)(nc_1 - c_2 - c_1)} - D(c_1 + c_2) \tag{14}$$

$$n^* = \begin{cases} \lceil n \rceil, \text{ if } \pi_C(\lceil n \rceil) > \pi_C(\lfloor n \rfloor); \\ \\ \lfloor n \rfloor, \text{ otherwise.} \end{cases} \tag{15}$$

$$Q_2^* = \sqrt{\frac{2D(\frac{K_1}{n^*} + K_2)}{I(n^* c_1 - c_2 - c_1)}} \tag{16}$$

$$Q_1^* = n^* Q_2^* \tag{17}$$

$\pi_C(n^*)$ is the optimal total profit of the centralized inventory system.

A numerical experiment is used to analyze the coordination failure, and the parameters in Eqs. 1–9 are valued from [4] as in Table 1.

Table 1. The data of known parameters

D	c_1	K_1	K_2	I
kg/day	$/kg	$	$	1%/year
189	30	700	120	30

The profits and optimal decisions of all firms of supply chain are calculated as in Table 2.

From Table 2, the node 2's profit π_2 and supply chain's profit, π_D, are all increased, as the unit wholesale price, c_2, is decreased, on the other hand the node 1's profit, π_1, is decreased. Therefore, there is not positive incentive to the node 1 to decrease its unit wholesale price for the node 2. That is to say, the supply chain locates at a suboptimal state of coordination failure (Table 3).

Table 2. The order quantities and performances for every c_2 (Decentralized)

c_2 ($/kg)	33	43	53	63	73	83
Q_2^* (kg)	1293.20	1132.89	1020.43	935.95	869.48	815.42
n^* (kg)	3	3	3	4	4	4
π_1 ($/day)	501.01	2390.14	4278.62	6167.04	8056.80	9946.28
π_2 ($/day)	12627.92	10732.96	8838.55	6944.54	5050.83	3157.37
π_D ($/day)	13128.94	13123.10	13117.17	13111.58	13107.63	13103.65

Table 3. The order quantities and performances for every c_2 (Centralized)

c_2 ($/kg)	33	43	53	63	73	83
n^*	1	2	2	3	3	3
Q_2^* (kg)	3380.51	1720.76	1613.77	1149.40	1105.35	1066.00
π_C ($/day)	13138.31	13126.75	13119.91	13113.80.01	13109.17	13104.71

4   Coordinating Policy by Revenue Sharing Contract

An approach to make both firms better off is to design terms of trade to give firms the proper incentive to choose supply chain optimal actions. If total supply chain profit increases, the "pie" increases and everyone can be given a bigger piece (Figs. 4 and 5).

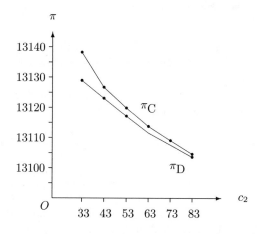

Fig. 4. The comparison of systematic profits in decentralized and centralized two-level inventory system

Table 4. The order quantities and performances for every c_2 and α (Decentralized, with Revenue Share Policy)

c_2 ($/kg)	33	33	33
α (%)	25	32	40
Q_2^* (kg)	3380.51	3380.51	3380.51
n^*	1	1	1
π_1 ($/day)	5252.86	6569.16	8087.86
π_2 ($/day)	7885.45	6569.15	5050.45
$\pi_1 + \pi_2$ ($/day)	13138.31	13138.31	13138.31

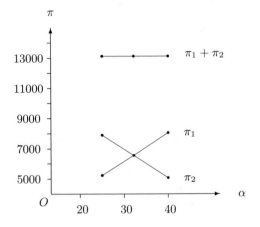

Fig. 5. The variable profits of system and subsystem in decentralized two-level inventory system with revenue share policy

In this section, we investigate how to coordinate supply chain via revenue sharing contract [5,8,9]. With revenue sharing, the node 1 as above can justify giving lower wholesale price.

Suppose the share of revenue of the node 1 as α, the profit functions of the node 1 and the node 2 are expressed as below

$$\pi_1(n, \alpha) = \alpha Dp + D(c_2 - c_1) - \frac{K_1 D}{nQ_2} - \frac{n-1}{2}Q_2 I c_1 \tag{18}$$

$$\pi_2(Q_2, \alpha) = (1 - \alpha)Dp - Dc_2 - \frac{1}{2}Q_2 I c_2 - \frac{K_2 D}{Q_2} \tag{19}$$

For certain given wholesale price c_2 and share of revenue α, the potential allocations of profit are estimated as in Table 4.

As α is increased, profit is shifted from the node 2 to the node 1. With a negotiation, there is an optimal combination (c_2, α) that supply chain reaches a global optimization.

5   Conclusion

The supply chain system is suffer from the coordination failure even though in the deterministic environment. The numerical experiment of simple two-level decentralized inventory system analyzes deeply the variation of the order quantities and system performances. The result shows that the revenue sharing contract can be used as a coordinating tool to make firms and the supply chain to global optimization.

Acknowledgment. The author would like to thank the Logistics and Supply Chain Management Discipline Cultivating Program of Shanghai Polytechnic University (XXKPY1606).

References

1. Arrow, K., Harris, T., Marschak, J.: Optimal inventory policy. Econometrica **19**, 50–272 (1951)
2. Gurnani, H., Drezner, Z.: Deterministic hierarchical substitution inventory models. J. Oper. Res. Soc. **51**, 129–133 (2000)
3. Andersson, J., Marklund, J.: Decentralized inventory control in a two-level distribution system. Eur. J. Oper. Res. **127**, 483–506 (2000)
4. Zhao, X., Huang, S.: Inventory Management. Tsinghua University Press, Beijing (2008)
5. Popiuc, M.N., Govindan, K.: Reverse supply chain coordination by revenue sharing contract: a case for the personal computers industry. EJOR **233**, 326–336 (2014)
6. Haan, M.: The economics of free internet access. JITE **157**, 359–379 (2001)
7. Gerard, C., Christian, T.: Matching Supply with Demand: An Introduction to operations Management, 3rd edn. McGraw Hill Higher Education, Beijing (2012)
8. Xiao, T., Xu, T.: Coordinating price and service level decisions for a supply chain with deteriorating item under vendor management inventory. Int. J. Prod. Econ. **145**, 743–752 (2013)
9. Berman, O., Crass, D.M., Tajbakhsh, M.: A coordinated location-inventory model. Eur. J. Oper. Res. **217**, 500–508 (2012)
10. Ma, H.: Game Theory. Tongji Press, Shanghai (2015)

Remote Intelligent Support Architecture for Ground Equipment Control in Space Launch Sites

Litian Xiao[1,2(✉)], Mengyuan Li[1], Kewen Hou[1], Fei Wang[1], and Yuliang Li[1]

[1] Beijing Special Engineering Design and Research Institute,
Beijing 100028, China
xiao_litian@sina.com
[2] Software School, Tsinghua University, Beijing 100084, China

Abstract. The control system of space launch site plays a key role in mission success. The application of intelligent control systems and unmanned systems can increase mission efficiency. However, these systems can also cause large uncontrollable security risks. Normally, in a launch site, experts must rush to the launch field for diagnosing and/or handling automatic control system faults. In the case of high-frequency multisite missions, this mode of operation is not sustainable. This paper proposes a remote and intelligent support architecture for the autonomous control of ground equipment. The architecture has three hierarchical levels: field, launch-site, and long-range. It integrates the control verification and fault diagnosis of field autonomous control at the launch site with remote fault diagnosis and prediction and remote maintenance. It enables the control system for the ground equipment in the launch site to operate without faults. This paper introduces the main schemes necessary to realize the key elements of the system and the architecture. In the preliminary practice, the launch mission is found to be supported effectively.

Keywords: Space launch site · Ground equipment control · Remote support · Intelligent architecture

1 Introduction

The conventional ground equipment control systems for space launch sites have centralized or distributed controls. The design, construction, and operation of such control systems are the responsibilities of different professionals. Conventional control systems do not have intelligent functions and are concentrated in the launch site field. In the case of a fault, the mission crew at all levels and experts will be dispatched to the site for handling, and the launch mission will be paused for system fault diagnosis and troubleshooting. The safety, timeliness, and success of space launch missions may be affected greatly by such fault events. In addition, the long response time for on-field troubleshooting will also affect the mission window and create mission losses [1].

In order to reduce the mission crew size at the launch site, now many ground facilities and equipment are under autonomous control. We propose and have designed

© Springer Nature Switzerland AG 2020
J. H. Kim et al. (Eds.): ICHSA 2019, AISC 1063, pp. 226–234, 2020.
https://doi.org/10.1007/978-3-030-31967-0_26

a remote intelligent support architecture using which the autonomous control system can locate and isolate faults in time. The automatic control system is designed to be lightweight in the field. The architecture simplifies the field control system and strengthens the remote intelligent support. It achieves front-end autonomous control and remote state detection and verification. Once field-state abnormalities or problems are detected, the intelligent support system can troubleshoot the faults and locate the fault points in a timely manner, or serve as a tool for expert consultation. The remote intelligent support architecture is designed in a hierarchical manner for intelligent control, supervision/monitoring, and management and support of ground equipment at the launch sites.

2 Related Research and Requirements

2.1 Control System at Launch Site

A general control system for the launch site can be described by a state-transfer model [2–4]. The autonomous control system of launch site completes the control task autonomously, according to the control logic and rules of a mission. The main fault types of programmable logic controller (PLC) based control systems are input faults (operation errors), sensor faults, actuator faults, and PLC software faults. PLC software faults can be avoided as far as possible through tests and verifications conducted in advance [5].

2.2 Intelligent Fault Diagnosis System

According to the results of the statistical analysis of ground facilities and equipment faults at the launch site, approximately 42% of the faults are difficult to detect through daily maintenance. The fault-detection measures for ground equipment are simple, and there are no criteria or means for quick fault detection. At present, research has achieved good results in terms of intelligent fault detection and diagnosis technologies for control systems [6, 7]. The three schemes are based on the analytical model, knowledge-based method, and signal-processing method, respectively. All these techniques can be applied to the fault detection and diagnosis of the control systems at the launch site. Their focus is on the diagnosis and isolation of faults using inferences based on prior knowledge and self-learning.

2.3 Technical Support

The general technical support system is designed to ensure the effective operation of the system. It usually includes the server terminal and the client terminal. By performing a series of operations on the server terminal, technicians can choose basic operation services, according to the application patterns and requirements of each operating platform. The client terminal serves as a human–computer interface for users [8]. At present, the technical support system mostly provides services on the website and does not have an intelligent service function.

2.4   Current Status of Control System at Launch Site

The control system at the launch site has gradually started providing unattended autonomous control during the flight mission. Because of the limited resources of the field control systems, verifying the accuracy of autonomous control is not completely possible, which may result in security risks. Daily maintenance and detection still need personnel.

The conventional control mode of the ground equipment at the launch site is a type of chain control, from the upper level computer to the PLCs driving the field actuators and motion devices [9]. Some manufacturers supply a central control unit to control the actuators and motion devices via Ethernet [10]. These control systems are still inadequate in terms of information management, analysis, and diagnosis.

In the case of a fault, technicians are assigned to resolve the problems at the launch site. This mode consumes many resources and involves high costs. Moreover, it will delay the mission. Now, if there are problems at multiple launch sites, it will be impossible to cope with multiple situations, simultaneously, especially, if there are frequent missions. Moreover, the same fault may occur at different launch sites, and fault only is manually detected and inferenced other cases.

3   Hierarchical Remote Intelligent Support Architecture for Ground Equipment

The control system for ground equipment plays an important role in ensuring accurate and reliable control of each system, and in completing the test preparation, propellant refueling, and launching.

Intelligent support must make system-performance forecasts when the critical equipment control performance has changed. It must have simple and convenient maintenance procedures and must ensure that the system is always in a good condition. In the event of a sudden control fault, it must make a rapid and accurate diagnosis, and based on the location, provide countermeasures on how to handle the situation appropriately. It must maintain control safety and reliability during the mission. These requirements can be met by constructing an intelligent support architecture with remote monitoring, control verification, fault diagnosis, and prediction. Intelligent support needs to provide work capabilities under multi-mission and multi-field scenarios.

The hierarchical remote intelligent support architecture for the ground equipment at the launch site is shown in Fig. 1. It supplies a platform for the launch service, to all professionals, including field and rear experts. The architecture is divided into three levels: field level, launch-site level, and long-range level.

3.1   Field Level

The control system for ground equipment is distributed at the launch site. Usually, the control system completes the control tasks independently, according to the task flow. Operators can operate the system through close control. In the case of intelligent autonomous control, the field will consume more computing resources. Field verification

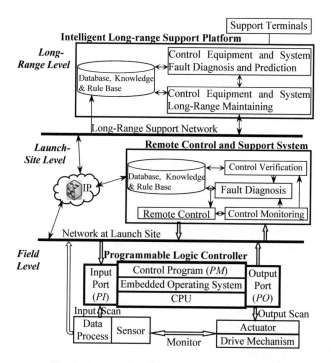

Fig. 1. Remote intelligent support architecture.

of whether the autonomous control is correct or not requires the deployment of additional resources. These will cause field-resource strains and inconveniences in terms of the control equipment layout and maintenance. Field control systems become more complex. Therefore, the front-end of the automatic control system is set up to be lightweight, to simplify the field control system.

In practice, data representation is mainly divided into two methods: feature-based and relation-based. The former describes the statistical information of data. The latter describes the entity association relationship in data [11]. Feature selection technology screens important features from high-dimensional space to reduce data dimension. The approach not only alleviates dimensional disaster but also compress big data [12].

3.2   Launch-Site Level

This level is located at the command center of the launch site. Intelligent verification verifies whether the front intelligent control is operating correctly or not. In the case of faults, the fault-diagnosis system automatically performs the diagnosis. Operators, technicians, and supervisors monitor the performance state of ground equipment in real-time, who can maintain system equipment and control programs. The level performs remote control in the case of abnormal situations and coordinates with remote experts for technical support and guidance. Thus, it enhances the remote rear-intelligence support.

3.3   Long-Range Level

The level provides a platform to comprehensively analyze the system status information and field-level data from the monitoring and collecting processes. Long-range technical support can be synchronized with the field during the mission. It provides intelligent technical consultations, troubleshooting expert consultations, equipment maintenance, early fault warning, maintenance guidance, and other services to the field-equipment operators.

3.4   Intelligent Support Flow in Mission

The hierarchical remote intelligent support system is briefly described as follows:

a. The field level performs autonomous control.
b. The launch-site level is for autonomous control verification and fault diagnosis.
c. The long-range level provides intelligent technical support for front-end autonomous control faults and maintenance.

The intelligent support processes involved in the control task are shown in Fig. 2.

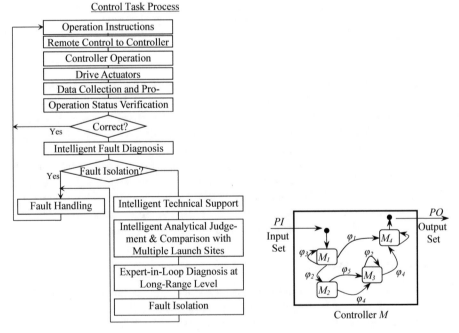

Fig. 2. Intelligent support processes in control task.   **Fig. 3.** State transferring diagram of controller

4   Hierarchical Remote Intelligent Support Architecture for Ground Equipment

There are three key elements in the proposed architecture. State verification judges whether the system is operating correctly or not. Intelligent fault diagnosis (IFD) ensures fault isolation and accurate handling, according to historical and prior knowledge [13]. Long-range technical support is used when the launch-site fault diagnosis fails to isolate some fault and undertakes daily maintenance and fault prediction.

4.1   State Verification of Remote Control

Because the unattended remote autonomous control is implemented on the field of the launch site, whether the execution of control instructions is correct and whether the control system performs normally in necessary to be verified. The core is the state verification of remote control.

Verifying the trust of controller state transfer is the key to determine whether the control system is running correctly or not [13]. Let us assume that there are n states inside a controller. Each state transformation is decided by M, which is constructed with $\{M_1, M_2, \ldots M_L\}$ inner states. The system involves multiple state transfers, labeled as $\varphi_i (i = 1, 2, \ldots n)$. Each M_1 may have several φ_i. The combined state transformations of multiple controllers are the same as inner states of individual controller. An example is shown in Fig. 3. Thus, the trusted elements of the controller include the input/output interface, configuration transformation, state transfer, and programmable components. In addition to the output set containing the operation drivers, M and φ_i variables are used for state verification. The correctness of the state transfer is verified by the controller in the launch-site level.

The model M used in the example consists of four submodules $M_i (i = 1, 2, 3, 4)$. $\varphi = \{\varphi_1, \varphi_2, \varphi_3, \varphi_4\}$ is a verified property, where $\varphi_i (i = 1, 2, \ldots 4)$ is the state-transfer property, from the input set to the output set. It must be ensured that the transfer property is trusted, from the input to the output. A concrete verification method should adopt the approaches of combination verification [14]. The PLC state data and output set are transmitted to the launch-site level through the network interface, and the launch-site level performs verification calculations.

4.2   Intelligent Fault Diagnosis

IFD is performed at both the launch-site and long-range levels. One function of IFD is to monitor the input and output of the field control on the basis of state verification. Another function is the automatic analysis and diagnosis of control faults after abnormal conditions are found. By adopting the technology of fault self-diagnosis and online fault diagnosis, the detection and consultation of field troubleshooting for software and hardware can be supported timely. Fault analysis and diagnosis include fault detection, fault isolation, and fault handling [15].

In the proposed architecture, the inference engine of IFD and analysis is mainly composed of the knowledge base, database, rule base, and interpretive modules. The IFD makes decisions and detections, given a threshold of the field information. Alternatively, the IFD process is started by a state-verification error. If a given threshold of the field information exceeds the fixed limits or a state-verification error occurs, it is probable that a fault occurs and the inference engine starts operating. Based on the field data and control tactics, IFD chooses the rules in the rule bases. Then, the system derives the diagnosis results using reasoning and rule matching. The fault location, cause analysis, and removing faulty programs are performed based on the results.

The control tactics adopted are different in the matching operation. When a fault conclusion is false, the application of the related rules is not given up at once. The reasoning is backtracked to a place recorded for the continuation of the next conclusion processing by the rules. Whether the relative rules can be used or not, is determined after the relative conclusions have been processed. Because matching reasoning rules can be allowed to pass through the matching operation, the inference process can be guaranteed when field information is not complete. This method is very important to improve the reliability of fault diagnosis. The inference engine derives results based on a mapping function that locates the relative knowledge by means of a return-to-one matching.

4.3   Intelligent Support

Fault analysis and diagnosis include fault detection, fault isolation, and fault handling [16]. One application of the proposed architecture in the Intelligent Long-range Support Platform (ILSP) is for control equipment and system fault diagnosis and prediction. The application functions have mainly four parts.

Real-time monitoring and early warning: The security of the control system can be mastered in real time. The security function can provide early warning about possible influences on the stable state of a control system and produce a correct response strategy to eliminate these influences as soon as possible. In the supervision and control of operation states, the state data of the electromechanical equipment controlled in the field are monitored.

Fault diagnosis and prediction: IFD is based on received operational data. A possible fault is predicted using historical data, knowledge data, and comparison of multiple launch sites. The operation and test data of electromechanical equipment are collected and stored by the intelligent support interface in real-time.

Maintenance: It includes the automatic control system applications of state prediction, maintenance planning, health analysis, and assessment. After obtaining authorization from the launch site, remote maintenance of the on-field control equipment can be carried out. The real-time health status of field equipment can be obtained by using the constructed control equipment health model and system principle model.

Security check: It not only manages the three-level user identity and authorization but also checks the control processes of the launch site.

5 Construction of Remote Intelligent Support Architecture

The data service of remote intelligent support is the basis of supervisory, control verification, fault diagnosis and prediction. The remote intelligent support platform provides data services to different applications in the form of cloud computing data centers and customized cloud service modes [3]. Database, knowledge base, and rule base adopt distributed data services on launch-site level and long-range level.

The construction of remote intelligent support architecture is based on the Internet of Things (IoT) of the launch site. The IoT application implementation uses object linking and embedding for process control unified architecture (OPC UA). A second generation OPC (object linking and embedding for process control) technology is used to implement cross-platform and web-service-based architecture [15–17]. In practical applications, OPC UA realizes information sharing between launch-site systems. It also connects field equipment, launch control execution systems (LCES), human–machine interface (HMI) systems, supervisory control and data acquisition (SCADA), and batch control (Batch). Meanwhile, it unifies the internal information exchange protocols in the remote intelligent support architecture, which reduces the operation and maintenance costs [17]. The developed middleware are mainly used for data interactions with various systems, releasing and sharing application services, acquisition and distribution of data information, and provision and invocation of interfaces.

6 Conclusion

We proposed a remote intelligent support architecture for ground equipment control at space launch sites, and we built and deployed this architecture at four launch sites in China. Using the ILSP constructed in Beijing, the previous procedure involving experts rushing to the launch site for providing technical support during each mission could be changed. Expert teams could support multiple launch sites simultaneously. The platform was used to handle control-system faults in several missions and to effectively ensure smooth completion of the launch missions.

In the cases where automatic fault diagnosis was not perfect, the expert-in-loop mode of remote intelligent support was quite effective. Since the application time of remote intelligent support is short, it is necessary to further enrich the diagnostic data, fault prediction, and health assessment models.

This paper introduced the overall framework of the architecture and its key elements. However, full details were not discussed here because of length limitations. Although the research needs further improvements, we believe that it has a good foundation [1–3, 12, 14]. Applications on the architecture can be connected through the middleware of IoT [14] and gradually extended to intelligent management and support of all facilities and equipment at launch sites.

References

1. Xiao, L.T., et al.: A hierarchical framework for intelligent launch site system. In: 2018 IEEE International Conference on Information and Automation, 5 p. IEEE Press, China (2018)
2. Xiao, L.T., et al.: System architecture and construction approach for intelligent space launch site. Adv. Intell. Syst. Comput. **856**(1), 397–404 (2019)
3. Xiao, L.T., et al.: Intelligent architecture and hybrid model of ground and launch system for advanced launch site. In: 2019 IEEE Aerospace Conference, 10 p. IEEE Press, USA (2018)
4. Xiao, L.T., et al.: PLC programs' checking method and strategy based on module state transfer. In: 2015 IEEE International Conference on Information and Automation, 5 p. IEEE, China (2015)
5. Arup, G., et al.: FBMTP: an automated fault and behavioral anomaly detection and isolation tool for PLC-controlled manufacturing systems. IEEE Trans. Syst. Man Cybern. Syst. **47**(12), 3397–3417 (2017)
6. Liu, W., et al.: Distributed intelligent fault diagnosis system based on system fusion. In: 2018 IEEE Aerospace Conference, 5 p. IEEE, USA (2018)
7. Xu, F., et al.: Robust MPC for actuator-fault tolerance using set-based passive fault detection and active fault isolation. J. Appl. Math. Comput. Sci. **27**(1), 43–61 (2017)
8. Victor, B., et al.: Concept implementation of decision support software for the risk management of complex technical system. Adv. Intell. Syst. Comput. **512**, 255–269 (2017)
9. Walker, M.S., et al.: Serial data machine control system. In: 16th Annual SMPTE Television Conference, 8 p. IEEE, USA (2015)
10. Liliam, R.G., et al.: Network-based control system to compensate the input delay and minimize energy expenditure of a cooling plant. In: 15th International Conference on Electrical Engineering, Computing Science and Automatic Control, 7 p. IEEE, USA (2018)
11. Shi, P.M., et al.: A novel intelligent fault diagnosis method of rotating machinery based on deep learning and PSOSVM. Adv. Intell. Syst. Comput. J. Vibroeng. **19**(8), 5932–5946 (2017)
12. Xiao, L.T., et al.: An approach for the verification of trusted operation on automatic control system. In: 2018 International Conference on Physics, Computing and Mathematical Modeling, 6 p. IOP, China (2018)
13. Marino, M., et al.: Distributed fault detection isolation and accommodation for homogeneous networked discrete-time linear systems. IEEE Trans. Autom. Control **62**(9), 4840–4847 (2017)
14. Xiao, L.T., et al.: An architecture of IoT application support system in launch site. Trans. Adv. Intell. Syst. Res. **863**, 169–173 (2018)
15. Garcia, M.V., et al.: OPC-UA communications integration using a CPPS architecture. In: 2016 IEEE Ecuador Technical Chapters Meeting, 7 p. IEEE, USA (2016)
16. Salvatore, C., et al.: Integrating OPC UA with web technologies to enhance interoperability. Comput. Stand. Interf. **61**, 45–64 (2019)
17. Alexander, G., et al.: TSN-enabled OPC UA in field devices. In: 2018 IEEE 23rd International Conference on Emerging Technologies and Factory Automation, 7 p. IEEE, USA (2018)

Correlation Analysis and Body-Weight Estimation of Yak Traits in Yushu Prefecture

Xiaofeng Qin[1], Yu-an Zhang[1(✉)], Meiyun Du[1], Rende Song[2],
Chenzhang[1], and Zijie Sun[1]

[1] Department of Computer Technology and Applications,
Qinghai University, Xining 810016, China
2011990029@qhu.edu.cn
[2] Yushu Prefecture Animal Husbandry and Veterinary Station,
Yushu 815000, Qinghai, China

Abstract. In this paper, body-related data of growth of 1011 healthy yaks as well as the data of their parents were collected from Qinghai Yushu farm. Through analysis it can be found that the yak grows very fast and their body change greatly from one-year-old to three years old, the average of body height, body length, chest girth, circumference of cannon bone and body weight increased percentage can reach 98.46%, 128.1%, 169.39%, 112.34%, and 115.66%. At the age of three, the traits are basically mature, and the yak with excellent traits can be feed as a breeding cattle while the yak with poor traits will be slaughtered. The excellent degree of parental traits has a significant effect on the trait characteristics of descendant yaks. Therefore, parental yaks must undergo strict screening in order to inherit the superior genes to the descendant yaks to a greater extent, which plays an important role in protecting good traits and improving breeding quality. The Linear Regression model was used to predict the weight of 3-years-old yaks and their parents, the results showed the model fitted well and the accuracy of training set reached 0.84, 0.99, 0.98 respectively, while the accuracy of test set reached 0.73, 0.97 and 0.99 respectively.

Keywords: Yak · Trait research · Body size · Body-weight estimation · Correlation analysis · Linear Regression prediction

1 Introduction

Yak is a peculiar primitive animal species in the Qinghai Tibet Plateau and its adjacent areas, which lives in areas with high altitude, low temperature, large temperature difference between day and night, short grass growing period, strong radiation and low oxygen partial pressure. Qinghai Plateau Yak is one of the main varieties of yak breeds, mainly distribute in 6 counties of Yushu Prefecture, 6 counties in Guoluo, 3 townships in the west of Xinghai County, Hainan Prefecture, Zeku in Huangnan Prefecture, 2 counties in Henan, Golmud Tanggula Township in Haixi Prefecture, etc. [1]. Yak has a very tenacious resistance, disease resistance, tolerance and adaptability, and it survives in the alpine meadow pasture where other livestock are difficult to live, which is not

© Springer Nature Switzerland AG 2020
J. H. Kim et al. (Eds.): ICHSA 2019, AISC 1063, pp. 235–243, 2020.
https://doi.org/10.1007/978-3-030-31967-0_27

only one of the indispensable milk sources in China, but also an indispensable resource for local herdsmen. The yak meat is tender and juicy, with delicious color and excellent taste. The milk is also rich in nutritional value.

At present, the analysis of body weight and body size of Qinghai yak indicates that there is a clear linear relationship between yak body weight and body size, the correlation between yak body size and body weight has been quantitatively analyzed and explained, and a multivariate linear regression equation of yak body weight and body size was established to describe the growth performance of yak [2–4]. Jie [5] used the path analysis method to calculate the direct and indirect effects of body weight traits on the body weight of yak, indicating that the direct effect of yak chest girth on body weight is greater than the effect of other traits, and the indirect effect of chest girth is on body height and body length has made a major contribution to the effects of body weight. Fan [6] studied the effects of different feeding patterns on the growth and development of juvenile yaks. By studying the changes in body size and body weight of yak found that the patterns conducive to growth and development, the use of cowshed management and feeding methods can accelerate the growth and development of young yaks. Zhao [7] studied the performance of Changtai Yak in the germplasm resources of different genders and age groups, which provided a data basis for the improved breeding of yak. Wu [8] pointed out that chest girth, body length, breast height and depth have important effects on the milk yield of yak when studying Maiwa yak and suggested that the above factors should be considered in yak breeding. Han [9] analyzed the body size and body weight of hybrid calves at different ages and genders found that the growth performance of hybrid improved yak was significantly improved, and the evaluation of the growth performance of hybrid yak was achieved. Based on the above research, it can be found that although it has a great contribution to the research and breeding of traditional yak, the results are relatively simple, and the amount of data set is very small. And lack the research of correlation analysis for yaks and their parents, which impossible to clarify the actual impact and changes brought about by the improvement and breeding of yak.

2   Materials and Methods

In this study, 1011 healthy breed yaks and their parents in Qinghai Province were selected as research objects and collected by professional researchers of Yushu Prefecture Animal Husbandry and Veterinary Station. The weighbridge accurately measures the body weight data of the yak in the fasting state, and uses the tape measure tools to measure the body height, body length, chest girth and circumference of cannon bone data of the yak. Recorded data of yaks from birth to three years old, each yak wears an immunological ear tag to uniquely mark the individual. The yak is divided into three levels by professionals according to yak body shape, body size, hair, reproductive organs, etc. The data is mainly described as:

Body weight (W): The yak was fasted after 12 h and weighed by the weighbridge in kilogram (kg);

Body length (BL): Also known as the lean body length, refers to the distance from the shoulder to the end of the ischial bone, measured with a tape measure in centimeter (cm);

Body height (H): the vertical distance between the highest point of the armor and ground contact point measured with a tape measure in centimeter (cm);

Chest girth (CG): the circumference of the vertical axis of the posterior scapula of the shoulder blade measured with a tape measure in centimeter (cm);

Circumference of cannon bone (CCB): the circumference of the thinnest part of the tube bone, which usually from the lower left anterior leg to the upper one third measured with a tape measure in centimeter (cm) (Fig. 1).

Fig. 1. The measurement figure of yak body size

3   Algorithm Implementation

3.1   Descriptive Analysis

The yaks from birth to less than one year old are recorded as 1-year-old(1), and those between one year and two years old are recorded as 2 years old(2), and those older than two years old and less than three years old are recorded as 3 years old(3). The H, BL, CG, CCB and W of the yak is analyzed and described separately, and the data shown in Table 1 can be obtained.

Table 1. Body size and weight change of yaks from birth to three years

Variables	Maximum	Minimum	Mean	Mode	Standard deviation
H(1)	82	38	53.41	51	2.66
BL(1)	65	47	50.71	49	2.31
CG(1)	70	49	56.98	60	3.25
W(1)	15	6	8.35	8	0.56
H(2)	18	13	15.15	15	0.85
BL(2)	139	91	101.69	101	2.21

(continued)

Table 1. (*continued*)

Variables	Maximum	Minimum	Mean	Mode	Standard deviation
CG(2)	142	103	107.09	137	4.40
CCB(2)	19	11	14.01	14	3.84
W(2)	156	106	138.38	140	0.78
H(3)	118	102	106	105	5.11
BL(3)	160	103	115.67	117	1.52
CG(3)	193	138	153.5	150	4.63
CCB(3)	19	11	17.73	18	0.56
W(3)	277	106	190.38	192	14.91

The life of yak is generally between 20 to 30 years old, the young yak begins to develop from about two years old. Male yaks exhibit sexual behavior at the age of one and reach mating age at 2 years old. It can be found that the yak grows very fast from the time of birth to the age of three by Table 1 (1, 2, 3 indicates the age of yak). From 1 to 2 years old, the average H can increase 48.28 cm, the percentage increase is 90.40%, the average BL increases 56.38 cm and the percentage increase is 111.18%, the average CG increases 78.88 cm and the percentage increase is 138.43%, the average CCB increases 5.66 cm and the percentage increase is 67.78%, the average W increases 123.23 kg and the percentage increase is 813.40%. From the age of 2 to 3, BH increases 4.31 cm and the percentage increase is 4.24%, BL increases 8.58 cm and the percentage increase is 8.01%, CG increases 17.64 cm and the percentage increase is 12.98%, W increases 52 kg and the percentage increase is 37.58%. After comparison, it can be found that the growth of yak from the first birth to the age of 2 is extremely rapid, and there is a huge change in both body size and body weight.

Considering the growth performance from the first birth to the age of 3, the average of H, BL, CG, CCB and W of the yaks can reach 98.46%, 128.10%, 169.39%, 112.33%, and 115.66% respectively. We can use standard deviation to measure the distribution of body weight and body weight of yak. The standard deviation of body size and body weight at the time of birth is small, which indicating that the body size and body weight of the newborn yak are basically at the same level, and the data distribution is relatively concentrated. The standard deviation of body length and body weight of 2-years-old yak has a minor increase, which indicating that the growth trend of yak at this stage has changed, body size and body weight increase rapidly for some yaks with good traits and body shape so that data dispersion increased. The 3-years-old yak has a significant improvement in the standard deviation of body weight indicating that the size and weight of the yak have been greatly different so that any yak without ideally growth state will not be considered as a breed yak and will be reared as ordinary yak.

Table 2. The body size and weight data of yak's biological father

Variables	Maximum	Minimum	Mean	Mode	Standard devition
Patenal Age	8	4	4.5	5	0.52
Patenal H	158	120	146	155	9.41
Patenal BL	151	107	133.52	137	4.73
Patenal CG	204	130	184.94	184	10.83
Patenal CCB	19	13	17.11	17	0.86
Patenal W	473	213	354.58	341	55.97

Table 2 shows the body size and weight data of the biological father of the yak. It can be seen that the male yak of 4-5 years old is more and densely distributed, and the weight of the yak reaches 524 kg that with excellent traits can undertake the work of genetically superior genes to the descendant yak. The CCB of a yaks has not changed much after being an adult. Therefore, the data has a dense distribution. Due to the different growth conditions of each yak lead to the H, BL, CG, CCB and W shows a big difference. There are 718 special grade breed yaks in this dataset, 290 at the first level and only 3 at the second level. The rating rules are directly related to the weight of yak, which is the reason that the discrete weight data.

Table 3. The body size and weight data of yak's biological mother

Variables	Maximum	Minimum	Mean	Mode	Standard deviation
Matenal Age	6	2	4	5	0.52
Matenal H	214	100	124.35	130	9.0
Matenal BL	177	103	113.99	117	6.1
Matenal CG	176	117	161.03	167	8.1
Matenal CCB	18	13	15.94	16	0.29
Matenal W	295	171	229.82	245	32.46

Similar to the male yak, the female yak of yak's mother have a similar distribution with the body weight data (Table 3). The weight of the female yak can reach 295 kg, which is quite different from the weight of the male yak. There are 242 individuals with a special grade, 516 at the first level and 253 at the second level, which is more stable than the extreme grade distribution of the male yaks.

Figures 2 and 3 shows the trend of W, H, BL, CG and CCB of yak from the first three years. It can be found that the weight has the maximum variation and significant increase, H, BL, and CG have consistent increase range. In the figure, the variation of the CG has a little change is mainly related to the improper scale of the graph. Normalization of the data can be found that the CCB is basically maintain the consistent increase with the other data.

Fig. 2. The trend of body size from 1–3   **Fig. 3.** The trend of body weight from 1–3

3.2   Correlation Analysis

Table 4 shows the Pearson correlation coefficient about the body size and body weight of yaks and their biological father that some potential rules are discovered: There is a significant negative correlation between the body height of a one-year-old yak and its parental body height, and it is negatively related to the paternal chest circumference and weight, but there is a significant positive correlation with the paternal tube circumference. There is a significant negative correlation between the body length of the yak and the paternal age and a significant negative correlation with the paternal tube circumference. The yak chest circumference and the parental age has a significant correlation at the 0.05 level. And the yak tube circumference with the body height, body length, chest girth and body weight has significant negative correlation. The weight of the yak is significantly correlated with the age, body size and weight of the parents.

Table 4. Correlation coefficient between one year old yak and its biological parents about body size and body weight

Variables	H	BL	CG	CCB	W
Paternal age	−.014	−.098**	−.072*	.030	−.074*
Paternal H	−.141**	−.004	.001	−.084**	−.213**
Paternal BL	−.080*	−.004	.008	−.085**	−.095**
Paternal CG	−.135**	.038	.053	−.127**	−.176**
Paternal CCB	.158**	.078**	.036	−.053	.087**
Paternal W	−.146**	.030	.040	−.124**	−.191**
Maternal age	.034	.053	.065*	.036	.111**
Maternal H	−.123**	−.025	−.019	−.079*	−.153**
Matenal BL	.039	.019	.005	−.030	.095**
Matenal CG	−.125**	.001	−.019	−.102**	−.146**
Matenal CCB	−.105**	−.042	−.051	−.098**	−.034
Matenal W	−.113**	.007	−.013	−.091**	−.124**

Note: ** significantly correlated at the 0.01 level (Bilateral),
* Significantly correlated at 0.05 level (Bilateral)

For the body size, body weight and marked level of the yak from 1 to 3 years old, the correlation coefficient in Table 5 can be obtained by correlation analysis. The level of yak is mainly related to the body size and the body weight at 3 years old and is the negative correlation, while the correlation is weak compared with the body size and weight at 1 or 2 years old. And the main reason is that the grading of yak is marked by manual, there is no accurate measurement of the rating (Table 6).

Table 5. Correlation coefficient between yak and age groups

Variables	H	BL	CG	CCB	W
Level 1	−.014	−.098**	−.072*	.030	−.074*
Level 2	−.056	−.155**	.024	−.021	.069*
Level 3	−.279**	−.342**	−.559**	−.133**	−.640**

Table 6. Spearman correlation coefficient between yak and parental level

Spearman correlation	Paternal level	Maternal level
Yak level	−.147**	−.044

The level of yak is expressed in discrete values, it cannot be measured using the common Pearson correlation to analyze the correlation between yak levels. Non-parametric statistics can be used to calculate the correlation coefficient based on the level of the rank rather than the actual value of the data that is Spearman rank correlation coefficient [10]. It is applicable to data that does not satisfy the normal distribution, grade data and data whose overall distribution is unknown.

$$r_s = 1 - \frac{6 \sum d^2}{n(n^2 - 1)}$$

Where d is the difference between the ranks of each pair of observations after the two variables X and Y are ranked, n is the number of all observation pairs. According to Spearman rank correlation analysis found that the correlation coefficient between the yak level and the father level is −0.147** which expressed in a significant correlation.

3.3 Linear Regression Prediction

The Linear regression(LR) is a data mining tool based on regression analysis in statistics and the quantitative function of regression analysis is one of the common methods in statistical analysis, the purpose is to find a line or a high-dimensional hyperplane that minimizes the error between the predicted value and the true value.

Regarded the body size of yak and the level as the independent variables, and the weight as the dependent variable to predict by LR algorithm, divided the training set and test set as the ratio of 4:1. The results of prediction is shown as Figs. 4, 5 and 6.

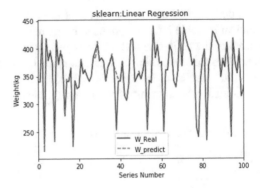

Fig. 4. The Comparison figure of real value and predicted value for three years old yak. The red line is the real value and green line indicates the predicted value. MSE (Mean Square Error) is 55.53, RMSE (Root Mean Square Error) is 7.45, training set accuracy is 0.84 and test set accuracy is 0.73.

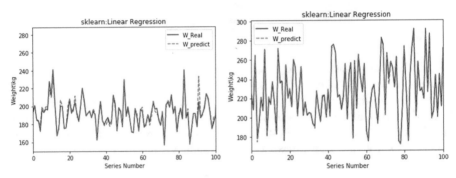

Fig. 5. The Comparison figure of real value and predicted value for yak's father. MSE is 82.93, RMSE is 9.11, training set accuracy is 0.99 and test set accuracy is 0.97.

Fig. 6. The Comparison figure of real value and predicted value for yak's mother. MSE is 13.28, RMSE is 3.64, training set accuracy is 0.98 and test set accuracy is 0.99.

4   Conclusion

In this paper, it is found that the growth rate of yak between 1 and 3 years old is extremely fast, especially between 1–2 years old. The average H, BL, CG, CCB and W increase can reach 111.18%, 90.40%, 138.43%, 87.78%, and 813.40% respectively. At the age of 3, the physical characteristics of the yak tend to mature, the yak with good traits is cultivated as a breeding yak while the yak with poor traits will be eliminated and raised as a normal yak for slaughtered. For the analysis of the parental yaks found that the traits of the male yaks are greatly different from each other so that the body size and the weight data are highly discrete. The female yak has a small difference in traits and the body size is less than male yak. In the correlation analysis, the trait characteristics of yak have a significant correlation with the traits of parents. Select the young, large-size body, high-grade breeders has a great influence on the traits of offspring. At the same time, the effect of the paternal yak on the little yak is more significant. Through the establishment of regression model using the yak body measurements and rating data to predict the weight, the training set accuracy of LR is 0.84, 0.99, 0.98 respectively, and the accuracy of test set is 0.73, 0.97, 0.99, which indicates the model has a good effect on the weight estimation. This study provides data support and theoretical basis for the breeding and research of yak in Qinghai Plateau.

References

1. Xian, G., Ping, Y.: Developmental situations and countermeasures of yak industry in China. China Cattle Sci. **35**(2), 55–57 (2009)
2. Li, H., Zhao, S.: Correlation analysis of body weight, body measurements in Datong yak. Chin. Qinghai J. Anim. Vet. Sci. **46**(4), 37–39 (2016)
3. Minqiang, W., Huiling, Zh.: Body weight growth model of Datong yak in Qinghai. China Herbivores **2005**(z2) (2005)
4. Qiong Da, L.: Correlation and regression analysis of the body weight and body size of adult female yak in Naqu. J. Anhui Agri. Sci. **39**(16), 9715–9716 (2011)
5. Jie, P., Ping, Y.: Multiple linear regression and path analysis between body measurement and weight on Datong yak. Chin. Herbivore Sci. **37**(6), 9–13 (2017)
6. Fan, F., Luo, Z.: Effects of different feeding patterns on growth and development of young yaks. Today's Anim. Husbandry Vet. **2018**(07), 44–45 (2018)
7. Zhao, H., Luo, X.: Study on the growth and development performance of Changtai yaks. Heilongjiang Anim. Sci. Vet. Med. **12**, 190–192 (2016)
8. Wu, J., He, S., Ai, Y.: Correlation between body size and breast traits and milk yield of maiwa yak and principal component analysis. Heilongjiang Anim. Sci. Vet. Med. **2019**, 149–152 (2019)
9. Han, X., Yan, Z., Jin, Y.: Effect of yak hybridization on its growth performance. Heilongjiang Anim. Sci. Vet. Med. **22**, 222–223 (2018)
10. Gauthier, T.D.: Detecting trends using spearmans rank correlation coefficient. Environ. Forensics **2001**(2), 359–362 (2001)

The Macroeconomic Influence of China Futures Market: A GARCH-MIDAS Approach

Ruobing Liu[1], Jianhui Yang[1], and Chuanyang Ruan[2,3(✉)]

[1] School of Business Administration, South China University of Technology,
Guangzhou 510640, China
[2] School of Business Administration,
Guangdong University of Finance and Economics, Guangzhou 510320, China
ruancyang@163.com
[3] Antai College of Economics and Management, Shanghai Jiao Tong Univerisity,
Shanghai 200240, China

Abstract. We revisit the relationship between the commodities futures market volatility and the macroeconomic factors, by employing the GARCH-MIDAS model, which can decompose the conditional variance into the secular and short-run component. We introduce the level or the variance of the macroeconomic variables into the GARCH-MIDAS model, to test the impact of the macroeconomic variables on the long-run variance. In the paper, we find the variance of PPI and IP has a more significant impact on the volatility of China commodities futures market than the level of the macroeconomic variables.

Keywords: GARCH-MIDAS · China futures market · Macroeconomic fundamentals · Long-run variance

1 Introduction

Macroeconomic news and volatility are known as the essential driver in the financial market. Many studies have researched studied the impact of the macroeconomic announcement on the stock market and commodity market. How the macroeconomic fundamentals influence the futures market, including the commodity futures and index futures? Previous studies investigated the impact of the macroeconomic fundamentals on the commodities and commodities futures. Our paper is to investigate the Chinese futures prices reaction of commodities futures and index futures to the macroeconomic fundamentals.

The previous study (see Zecchin et al. 2011; Zhu and Wang 2010; Ren et al. 2016; Teo et al. 2001) focuses on the impact of the macroeconomics, and the relationship between the volatility of the futures prices and the macroeconomic variables. The paper study the relationship between the macroeconomic fundamentals and futures prices volatility in China's futures market. Three essential questions are addressed in the paper.

Supported by Science and Technology Planning Project of Guangdong Province, China, (Grant No. 2014B080807027).

J. H. Kim et al. (Eds.): ICHSA 2019, AISC 1063, pp. 244–251, 2020.
https://doi.org/10.1007/978-3-030-31967-0_28

First, what is the impact of the macroeconomic variables on the volatility of the futures prices? Second, does the macroeconomic variables impact the commodities futures and the index futures in the same way? Third, how to measure the impact of the macroeconomic variables? The answers to these questions are essential for several aspects. First, the research is based on the daily data of the futures data, which allow us to explore the futures market reaction to the macroeconomic changes. In this research, it is essential to provide empirical evidence which illustrates the link between macroeconomic announce and futures price changes. Besides, an investigation of how the index futures responds to the macroeconomic news provide meaningful research. Our study provides a comprehensive study in the evaluation of the futures markets. The study not only includes the commodity futures but the index futures. We employ the GARCH-MIDAS model to construct the long-run and short-run variance in China's futures market.

In the research, we employ the GARCH-MIDAS model to solve the data mismatching frequency of daily returns and monthly or quarterly macroeconomic variables. The GARCH-MIDAS approach proposed by Engle et al. (2013). Based on the seniors' research, Engle et al. (2013) revisit modeling the economic sources of volatility, proposing a model which specify with the direct links to economic activity. The new model is called GARCH-MIDAS model. This model employs the mean-reverting daily GARCH process, and the use of the MIDAS approach is used to link the macroeconomic variables to the long-term variance. The long-term variance described by the MIDAS filter has the long memory features of the volatility process (Engle and Rangel 2008). The GARCH-MIDAS model can decompose the conditional variance into the short-run and long-run variance, the short-run variance is decided by the mean-reverting GARCH(1,1) process, the secular component of the variance is determined by the history of the realized variance or macroeconomic variables weighted MIDAS polynomials.

This model allows us to link the macroeconomic variables and futures volatility. It can also test the relationship between macroeconomic with different frequency and the specification of the long-run component. The GARCH-MIDAS model is widely on the research of the financial market. Conrad and Loch (2015) use the GARCH-MIDAS model to explore the relationship between macroeconomic fundamentals and the U.S. stock market volatility. Pan and Liu (2017) employ the GARCH-MIDAS model for the leverage effect in the short-run and long-run component. Mo et al. (2018) and Fang et al. (2018) use the GARCH-MIDAS model to explore the impact of the macroeconomic fundamental on the emerging commodities futures market. Donmez and Magrini (2013) study the macroeconomic determinants of the agricultural commodity price volatility by employing GARCH-MIDAS model.

The remainder of the paper is organized as follows: Sect. 2 is data, which is about the Chinese commodities futures market data and the macroeconomic variables data. Section 3 describes the GARCH-MIDAS model. Section 4 covers the empirical implement. The last section is the conclusion of the paper.

2   Data and Methodology

2.1   Data

According to the previous research, the macroeconomic variables are employed in the paper: inflation rate (CPI), Producer Price Index (PPI), Industrial Production (IP), Money supply growth rate (M2), Short-term interest rate. The inflation variable (CPI) measures the changes in the price level of the market basket of consumer goods and services purchased by consumers. This index measures the declines in the value of money in the economy. Close to the CPI, PPI measures the average changes in prices received by domestic producers for their output; this index measures the price changes from the viewpoint of the producers. As the measure of the output of the industrial sector of economy, IP is a vital tool to forecast future economic performance. IP is a small portion of the GDP. The GDP is viewed as the world's most powerful statistical indicator of national development and progress. Money supply M2 is the measure of the money supply including the cash, checkpoint, deposits, money market securities, mutual funds, and other time deposits. The M2 is considered as an indicator of money supply and future inflation. Short-term interest rates stand for the rates at which short-term borrowing are effected between financial institution or the rate the government paper is issued. In the research, we employ China current deposit interest rate as the short-term interest rate.

In the paper, our empirical results are based on the daily data of the commodities futures and the stock index futures. For the commodities futures, the research period is from June 2002 to December 2017, the stock index futures' research period is from June 2010 to Dec 2017. The futures settlement price is provided by the DataStream. From the statistics summary of the commodities futures and CSI 300 futures, we can know the settlement prices of the copper and soybean futures are left-skewed distribution, which have a long-left tail. The means of the copper and soybean futures' settlement price are left to the peak. The settlement prices of the aluminum and CSI 300 are right-skewed distribution, which means the distributions have long right tails. The means of the aluminum and CSI 300 futures prices are right to the peak. From the kurtosis of the futures settlement prices, we know the distributions of the futures prices have a lighter tail than the normal distribution.

2.2   GARCH-MIDAS Model

Based on the work of various component models (see Ding and Granger 1996; Ghysels et al. (2006)), Engle et al. (2013) propose the GARCH-MIDAS model. The GARCH-MIDAS model can decompose the conditional variance into two components. The short-run variance is a mean-reverting GARCH(1, 1) process, the secular component of the conditional variance is determined by the history of the realized volatility or the macroeconomic variables weighted by the MIDAS polynomials. The GARCH-MIDAS model can link the volatility directly to economic activity.

Following Engle et al. (2013), the GARCH-MIDAS can be specified as follows:

$$r_{i,t} - \mu = \sqrt{\tau_i g_{i,t}} \xi_{i,t} \tag{1}$$

$$\xi_{i,t}|\Phi_{i-1,t} \sim N(0,1) \tag{2}$$

In the Eq. (1), the $r_{i,t}$ is the log return of the day i in month t, τ_i, $g_{i,t}$ are different components of volatility, $g_{i,t}$ stands for the short-term component of volatility, τ_i is the secular component which accounts for the monthly and weekly volatility. In the Eq. (2), $\Phi_{i-1,t}$ is the information set-up to day $i-1$ of period t.

Inspired by Engle and Rangel (2008), GARCH-MIDAS model assumes the volatility dynamics of the short-run volatility $g_{i,t}$ is a GARCH(1,1) process, which can be written as follows:

$$g_{i,t} = (1 - \alpha - \beta) + \alpha \frac{(r_{i-1,t} - \mu)^2}{\tau_i} + \beta g_{i-1,t} \tag{3}$$

The long-run component of the variance for the GARCH-MIDAS model based on the Merton (1980) and Schwert (1989). In the paper, we don't view the realized volatility of a single month or quarter as a measure of interest; we specify the secular component τ_t by smoothing realized volatility in the spirit of the MIDAS filter and MIDAS regression:

$$\tau_t = m + \theta_1 \sum_{k=1}^{K} \varphi_k(\omega_1, \omega_2) \, RV_{i-k} \tag{4}$$

In Eq. (4), RV_{i-k} stands for the realized volatility in time $i - k$. $RV_{i-k} = \sum_{j=1}^{N} r_{i-j}^2$, which i stands for the day in trading period.

In Eq. (4), we use different weight schemes for the realized volatility and macroeconomic variables. In GARCH-MIDAS model, we set the conditional variance as follows:

$$\sigma_{i,g}^2 = \tau_i \times g_{i,g} \tag{5}$$

We specify weight scheme in Eqs. (5) and (6), and describe the $\varphi_k(\omega_1, \omega_2)$ as *Beta*, and the $\varphi_k(\omega)$ as the *Exponentially weighted*, the polynomials can describe as below:

$$\varphi_k(\omega_1, \omega_2) = \frac{(\frac{k}{K})^{\omega_1 - 1}(1 - \frac{k}{K})\omega_2 - 1}{\sum_{j=1}^{K}(\frac{j}{K})\omega_1 - 1(1 - (\frac{j}{K})^{\omega_2 - 1})} \tag{6}$$

$$\varphi_k(\omega) = \frac{\omega^k}{\sum_{j=1}^{K} \omega^j} \tag{7}$$

The $\varphi_k(\omega_1, \omega_2)$ in Eq. (6) stands for the Beta, the $\varphi_k(\omega)$ in Eq. (7) is the Exponentially weighted, the sum of the exp weighted is 1. The Beta lag polynomial has a further discussion in Ghysels et al. (2006) paper, the Beta can be represented monotonically increasing or decreasing in the weighting scheme.

The Eqs. (1)–(7) form the GARCH-MIDAS model for time-varying conditional variance with fixed time span RVs and the parameter space $(\Theta = \mu, \alpha, \beta, m, \theta, \omega_1, \omega_2)$. Comparing to other models, GARCH-MIDAS model has its characteristics. First, the number of parameter space is fixed. Unlike other component model introduce by Engle and Rangel (2008), GARCH-MIDAS model use parsimonious parameters to describe the dynamic of the volatility. Second, as the number in the GARCH-MIDAS model is set, we can compare the GARCH-MIDAS models with different periods. We can vary t concerning the period covered by the realized variance. The GARCH-MIDAS model with the realized variance allows us to incorporate the macroeconomic time series directly. In the paper, we employ the GARCH-MIDAS models with one-sided filters, which contain the macroeconomic variables.

$$\tau_t = m_l + \theta_l \sum_{k=1}^{K_l} \varphi_k(\omega_{1,l}, \omega_{2,l}) X_{t-k}^{mv} \tag{8}$$

In the Eq. (8), X_{t-k}^m stands for the macroeconomic variable m at time $t - k$. GARCH-MIDAS model introduces the macroeconomic variables as the explanatory variables. This model has a direct relationship between the macroeconomic and the underlying asset price change.

3   Empirical Results

During selection procedure, we use the BIC (Bayesian Information Criterion) as the standard to select the optimum number of lags and the exponential weights. In the paper we set the period is 22, which are the monthly aggregation. We set the period is 22, and the number of the MIDAS lags is 9, which means a history of 9 months' realized volatility will be averaged by the MIDAS weights to determine the long-run conditional variance. The 9 lag months cost 198 observations for initialization. In the paper, we employ the GARCH-MIDAS model with fixed-span RV and the model with rolling window RV. For the GARCH with fixed-span RV, the long-run variance is set as below:

$$\tau_t = m + \theta \sum_{k=1}^{K} \varpi_k(\omega_1, \omega_2) RV_{t-k} \tag{9}$$

$$RV_t = \sum_{i=1}^{N_t} r_{imt}^2 \tag{10}$$

With the introduction of the macroeconomic, we can know For the Soybean futures market, the θ is negative, which means the increase of the PPI level will lead to the low Soybean stock market volatility. With the similar computations, we can see 0.0015% decrease in the market volatility in the current quarter when there was a 1% point increase in the inflation 15 months ago. Our sample period covers the 2008 financial crisis and the 2015 China stock market crash.

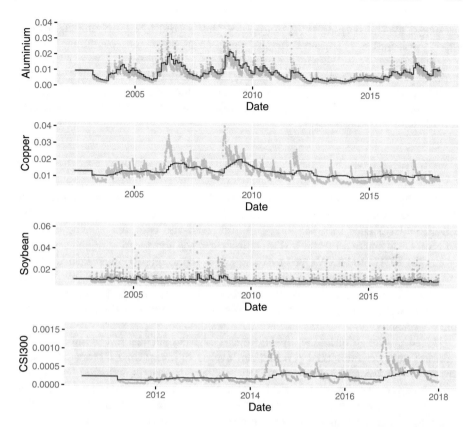

Fig. 1. The conditional variance and its secular component of the futures contract.

Based on the empirical result of the GARCH-MIDAS model, we can know the PPI level has an insignificant impact on three commodities futures markets (Fig. 1).

While for different commodities futures market, the influence of the PPI level is different, the increase of PPI level decreases the volatility of the in Copper and Soybean futures market, while the increase of the PPI level increases the volatility of Aluminum futures market. For the different commodities futures market, the impacts of the level of PPI and IP are different. For the Aluminum futures market, the increase of the PPI and IP level would increase the volatility of the Aluminum futures returns, while the increase of the PPI and IP level would decrease the volatility of the Copper futures and the Soybeans futures returns. After introducing the PPI variance or IP variance into the GARCH-MIDAS model, we can know the increase of inflation uncertainty and the industrial production uncertainty would increase the volatility in the three commodities market.

Based on the parameter estimates for the GARCH-MIDAS model, we can know the impacts of the level or the variance of the macroeconomic variables on the three commodities market are statistically insignificant. Even though the impacts are not significant, while for the three commodities market, the impact of the inflation uncertainty is more significant than the inflation level, and the impact of the industrial inflation uncertainty is more significant than the impact of industrial production level.

4   Conclusion

In this paper, we employ the GARCH-MIDAS model, which is the combination of the spline- GARCH and MIDAS filter. This model can decompose the conditional variance into the long-run and short-run component. The long-run component is decided by the MIDAS filter, and the short-run variance is decided by the GARCH model. We also link the macroeconomic variables with the long-run volatility in the commodities futures market. In the paper, we employ the GARCH-MIDAS model to test the influence of the current or previous macroeconomic variable change on the future volatility of the three commodities futures market returns.

Based on the empirical analysis, we can draw a conclusion: the impact of the inflation and the industrial production level is different for different commodities futures market, the increase of PPI level and industrial inflation level in the current or previous quarter would increase the future volatility of the Aluminum futures market returns, while the increase of inflation level and industrial inflation level in the previous or current quarter would decrease the future volatility of the Copper and Soybean futures market return. The increase of the uncertainty of inflation or the industrial production in the previous or current quarter would increase the market volatility next quarter.

References

Engle, R.F., Rangel, J.G.: The spline-GARCH model for low-frequency volatility and its global macroeconomic causes. Rev. Finan. Stud. **21**(3), 1187–1222 (2008). https://doi.org/10.1093/rfs/hhn004

Fang, L., Yu, H., Xiao, W.: Forecasting gold futures market volatility using macroeconomic variables in the United States. Econ. Model. **72**, 249–259 (2018). https://doi.org/10.1016/j.econmod.2018.02.003

Ghysels, E., Sinko, A., Valkanov, R.: MIDAS regressions: further results and new directions. Econometric Rev. **26**(1), 53–90 (2006). https://doi.org/10.1080/07474930600972467

Merton, R.C.: On estimating the expected return on the market. J. Finan. Econ. **8**, 323–361 (1980)

Mo, D., Gupta, R., Li, B., Singh, T.: The macroeconomic determinants of commodity futures volatility: evidence from Chinese and Indian markets. Econ. Model. **70**, 543–560 (2018). https://doi.org/10.1016/j.econmod.2017.08.032

Pan, Z., Liu, L.: Forecasting stock return volatility: a comparison between the roles of short-term and long-term leverage effects. Phys. A (2017). https://doi.org/10.1016/j.physa.2017.09.030

Schwert, G.W.: Why does stock market volatility change over time? J. Finan. **44**(5), 1115–1153 (1989)

Teo, K.K., Wang, L.P., Lin, Z.P.: Wavelet packet multi-layer perceptron for chaotic time series prediction: effects of weight initialization. In: Computational Science – ICCS 2001, Proceedings Pt 2, vol. 2074, pp. 310–317 (2001)

Ren, Y., Suganthan, P.N., Srikanth, N., Amaratunga, G.: Random vector functional link network for short-term electricity load demand forecasting. Inform. Sci. **367–368**, 1078–1093 (2016)

Zhu, M., Wang, L.P.: Intelligent trading using support vector regression and multilayer perceptrons optimized with genetic algorithms. In: International Joint Conference on Neural Networks (IJCNN 2010) (2010)

Zecchin, C., Facchinetti, A., Sparacino, G., De Nicolao, G., Cobelli, C.: A new neural network approach for short-term glucose prediction using continuous glucose monitoring time-series and meal information. In: Annual International Conference of the IEEE Engineering in Medicine and Biology Society, pp. 5653–5656 (2011)

Conrad, C., Loch, K.: The variance risk premium and fundamental uncertainty. Econ. Lett. **132**, 56–60 (2015). https://doi.org/10.1016/j.econlet.2015.04.006

Ding, Z., Granger, C.W.J.: Modeling volatility persistence of speculative returns: a new approach. J. Econom. **73**, 185–215 (1996)

Dönmez, A., Magrini, E.: Agricultural Commodity Price Volatility and Its Macroeconomic Determinants: A GARCH-MIDAS Approach. Working Paper (2013)

Engle, R.F., Ghysels, E., Sohn, B.: Stock market volatility and macroeconomic fundamentals. Rev. Econ. Stat. **95**, 776–797 (2013). https://doi.org/10.1162/REST_a_00300

Author Index

A
Ai, Longhai, 175
Al-Shamiri, Abobakr Khalil, 82

B
Bacanin, Nebojsa, 186
Bai, Kai, 111

C
Chenzhang, 235
Choi, Young Hwan, 70, 76, 82, 202

D
Dong, Jin-Wei, 160
Du, Meiyun, 235

F
Feng, Huabin, 148

G
Gao, Chong, 139
Gao, Hong Mei, 210
Gong, Shaoyan, 139

H
Han, Lu, 111
He, Zhenpeng, 130
Holmberg, Johan, 194
Hou, Kewen, 226
Huang, Geyu, 1
Huang, Weibo, 130
Huang, Wei-bo, 160
Hwang, Yoon Kwon, 63

J
Ji, Feng, 139
Jung, Donghwi, 76

K
Karamoddin, Negar, 202
Kim, Hyeong Suk, 169
Kim, Joong Hoon, 63, 70, 76, 82, 202
Kwon, Soon Ho, 63, 76

L
Lee, Chan Wook, 169
Lee, Eui Hoon, 63
Li, Mengyuan, 226
Li, Mingai, 54
Li, Tingting, 139
Li, Xiaodan, 130
Li, Xiao-Dan, 160
Li, Yong, 25
Li, Yuliang, 226
Li, Zheng, 44
Liao, Yan-Jun, 160
Lim, Oseong, 70
Lim, Seong Hyun, 169
Liu, Hengchang, 148
Liu, Jianze, 16
Liu, Kun, 44
Liu, Ruobing, 244
Liu, Shi, 175
Liu, Yu Chuan, 210

M
Ma, Li, 175
Min, Kyoung Won, 82

O
Ojha, A. K., 89

P
Pan, Meng, 35

Q
Qi, Jialin, 16
Qin, Xiaofeng, 235
Qiu, Shiyang, 148

R
Ramu Naidu, Y., 89
Ruan, Chuanyang, 244

S
Sadollah, Ali, 202
Shan, Yi, 54
Sheng, Jun, 16
Song, Rende, 235
Song, Shuaifei, 148
Strumberger, Ivana, 186
Sun, Yang, 175
Sun, Zijie, 235
Susheela Devi, V., 89

T
Tao, Zhengru, 111
Tuba, Eva, 186
Tuba, Milan, 186

W
Wang, Fei, 226
Wang, Yan, 16
Wu, Bizhu, 35
Wu, Jingyi, 122
Wu, Ming, 99
Wu, Suishuo, 54

X
Xiao, Litian, 226
Xiong, Ning, 194
Xu, Bingbing, 54
Xu, Peng, 148

Y
Yang, Jianhui, 244
Yang, Jinfu, 54
Yang, Jun, 99
Yang, Tao, 218
Yoo, Do Guen, 70, 169
Yu, LongQi, 16

Z
Zeng, Zixian, 25
Zhang, Jun, 1, 25
Zhang, Yangcheng, 16
Zhang, Yonghe, 35
Zhang, Yu-an, 235
Zhang, Zhe, 99
Zhang, Zhiming, 1
Zheng, Wei, 148
Zhou, Jianhui, 139